FAO Fisheries Report No. 230

FISH SILAGE PRODUCTION AND ITS USE

Edited by

J.G. Disney
Tropical Products Institute
London, U.K.

and

D. James
Fish Utilization and
Marketing Service
FAO
Rome, Italy

PAPERS PRESENTED

at the

INDO-PACIFIC FISHERIES COMMISSION

WORKSHOP ON FISH SILAGE PRODUCTION AND ITS USE

Jakarta, Indonesia, 17-21 September 1979

FOOD AND AGRICULTURE ORGANIZATION OF THE UNITED NATIONS

Rome, March 1980

The designations employed and the presentation of material in this publication do not imply the expression of any opinion whatsoever on the part of the Food and Agriculture Organization of the United Nations concerning the legal status of any country, territory, city or area or of its authorities, or concerning the delimitation of its frontiers or boundaries.

M-47

ISBN 92-5-100921-X

PREPARATION OF THE REPORT

The document contains all the papers presented at the IPFC Workshop on Fish Silage Production and its Use, which was held in conjunction with the Fourth Session of the IPFC Working Party on Fish Technology and Marketing, Jakarta, 17-21 September 1979. It also contains a comprehensive bibliography on all aspects of the production of fish silage and its use as animal feed. The editors have attempted to ensure that all of the references on silage quoted in the papers appear on the bibliography. Any inconsistencies result from material cited in the papers being at an early stage of preparation or not being generally available.

Distribution

FAO Fisheries Department
FAO Regional Fishery Officers
Authors
Participants
Institutes in the field of Fish Technology

Bibliographic entry:

Disney, J.G. and D. James (eds), Fish Silage 1980 production and its use. Papers presented at the Indo-Pacific Fisheries Commission Workshop on Fish Silage Production and its Use. Jakarta, Indonesia, 17-21 September, 1979. FAO Fish. Rep., (230):105 p.

TABLE OF CONTENTS

INTRODUCTION

The Indo-Pacific Fisheries Commission has appointed a number of working parties to investigate particular aspects of fisheries development. The Working Party on Fish Technology and Marketing was formed to foster collaboration between research institutes in the region and promote activities on the handling, processing and marketing of fish. An international symposium entitled Fish Utilization Technology and Marketing, was held in Manila, Philippines in March 1978.

At the third session of the working party, held in Kuala Trengganu in Malaysia, it was decided that the group should concentrate upon four broad areas, namely:

Fish silage
Improvement of salted and dried fish products
Storage of fish at various temperatures
Minced fish, utilization of by-catches

It was further agreed that in future the meeting of the working party would be extended to five days to permit the holding of a three-day technical workshop on one of the four topics covered in the overall work programme. Subsequently the decision was made to hold the fourth session of the working party in Indonesia in September 1979.

The fourth session was held at the Garden Hotel, Jakarta, Indonesia on 17 September 1979. Some 30 participants took part in the working party and the workshop on fish silage which followed. These proceedings report the outcome of the workshop on fish silage.

Before proceeding with the report of the workshop the relevance of fish silage to the IPFC region and the activities of the working party on this topic will be summarized.

For the preparation of fish silage, fish is ground or minced and mixed with acid (sulphuric, formic or propionic) to prevent bacterial putrefaction. An alternative process is to introduce lactic acid producing organisms to a mixture of carbohydrate and ground fish, the acid produced by fermentation then being responsible for the preservation. The proteolytic enzymes present in the fish cause the product to become more liquid, the rate of change being temperature-dependent. The liquid product may be fed direct to pigs or it may be mixed with a variety of agricultural products and sun-dried for feeding to poultry. A similar product may also be used as a fish food.

The fish silage process is simple and the capital investment required is relatively low when compared with fish meal production. Fish silage can also be produced on both a small and large scale and it therefore has potential application at the village level. The process could therefore have considerable potential in developing countries as a means of utilizing waste fish which cannot be used directly as human food. It is because the fish silage process could provide a more cost-effective means of using waste fish as animal feed that it was adopted as one of the programmes of the IPFC working party.

The fish silage process is not new. It was developed in Finland in the 1920s and taken up in Sweden, Denmark and Poland in the 1950s. In 1970 work was started in the United Kingdom and Australia and in 1974, as a result of collaboration between the CSIRO in Australia and TPI in London, silage was adopted as an IPFC work programme. Activity within the programme has resulted in work being started in Sri Lanka, India, Indonesia, New Zealand, Malaysia, Thailand, Hong Kong and the Philippines. These activities are being coordinated by the working party and several more specific programmes have been initiated. For example a Norwegian worker spent 6 months in Indonesia to carry out research on the production and feeding of fish silage. Two consultancies have also been carried out to assess the prospects for fish silage in countries within the IPFC region.

Therefore, there is considerable interest in fish silage in the IPFC region and it was entirely appropriate for the first workshop of the working party to consider the subject of fish silage.

The workshop attracted twenty papers although some were presented as abstracts only. A list of the papers is presented as Appendix A and a list of participants is given as Appendix B.

The workshop consisted of a review of the subject to cover areas such as the available resources, production, animal feeding trials, problems of commercial production, etc. The review included a wide ranging discussion and the presentation of an annotated bibliography on fish silage. This bibliography was expanded during the workshop and a revised version is presented at Appendix C. The bibliography represents a comprehensive coverage of all publications on fish silage. Appendix D provides annotations to the bibliography, while Appendix E is an author index to the papers in the Proceedings.

Following the review of the subject individual papers were presented. For ease of presentation the papers will not be presented in the same order as followed during the workshop. In fact the papers do not lend themselves to a clear separation into categories as several papers cover all aspects of silage from production to feeding and the economics of production. Nevertheless, the papers have been divided into the following categories although there may be some overlap:

The production of fish silage
The nutritive value of fish silage
Economic aspects

Points of discussion arising from the papers are not considered with the paper but an overall discussion is presented to include points of disagreement, areas requiring more work, recent findings, etc. This discussion is then followed by the conclusions and recommendations of the workshop.

THE PRODUCTION OF FISH SILAGE - PROSPECTS AND PROBLEMS

As indicated in the Introduction it was difficult to separate the papers into discrete subject areas but nine papers are presented in this section. These papers have the following titles:

POSSIBLE PROBLEMS ENCOUNTERED IN THE APPLICATION OF FISH SILAGE IN INDONESIA

R. Moelyanto
Research Institute of Fishery Technology
Jakarta, Indonesia

Abstract

The availability of raw material is one of the important factors for the production of fish silage. In Indonesia relatively cheap fish is available, both freshwater and marine. During the dry season freshwater fish is caught in quantity in open waters and at the height of the season for marine fish a lot of fish is available for silage production. Further sources of raw material for fish silage are by-catch from shrimp trawling and the offal from fish processing industries.

Although raw material is available its use for silage production is restricted by constraints such as the freshness of the raw material when it reaches the processor, government policy, availability and cost of acids, species of trashfish, transportation and storage of the silage and the possible occurrence of fishy taint.

Despite these problems and constraints research on the processing and the practical utilization of fish silage should be continued so that small-scale fishermen and fish farmers can benefit from the simple technology of fish silage production.

1. INTRODUCTION

In Indonesia, especially on Java which is the biggest market for fresh and cured fish products, all landed fish will be consumed or utilized. However, according to recent information and personal interviews with fishermen (which have to be verified) it seems that in the western part of the Indonesian archipelago some of the catch, which is not worth landing, is thrown back to the sea.

A rough calculation of fish waste indicates that some 225 000 tons of trashfish is thrown back to the sea. A further 200 000 tons of fish cannot be utilized due to spoilage or damage during handling, storage and distribution. Additional waste also occurs during the canning and freezing of shrimp (± 25 percent) and frog legs (± 70 percent). Thus, there is potentially a large source of raw material available for silage production in the region of 500 000 tons annually.

The main source of protein for animal feed in Indonesia is fish meal which is largely imported, and increases in the price of fish meal directly influence the price of the eggs and meat. The introduction of fish silage production could open up new possibilities for more efficient fish utilization and possibly provide a cheaper source of protein for animal feed.

2. SPECIES AND CHEMICAL COMPOSITION OF TRASHFISH

Preliminary research on the trashfish from the Java Sea which is landed in fairly fresh condition revealed the species composition, shown in Table 1. The chemical composition and amino acids of selected species is shown in Table 2. The total landings of trashfish in 1977 are given in Table 3 and the various means of utilizing the catch are illustrated in Table 4.

Table 1

Species composition of the by-catch caught by shrimp trawler in the Java Sea

Species	Percentage in the total catch	
Leiognathidae	24.9	
Leiognathus splendens		10.7
Mullidae	11.3	
Upeneus sulphureus		11.0
Carangidae	8.2	
Seloroides leptolepis		1.9
Gerreidae	7.1	
Pentaprion longimanus		6.9
Nemipteridae	6.6	
Nemipterus spp.		6.2
Scolopsis spp.		0.4
Priacanthidae	5.3	
Priacanthus spp.		5.3
Ariidae	4.8	
Arius thalassinus		2.7
Lutjanidae	4.7	
Lutjanus sanguineus		3.1
Synodontidae	3.8	
Trichyuridae	3.1	
Clupeidae	2.1	
Anadontostoma chacunda		0.1
Ilisha spp.		0.2
Balistidae	1.9	
Abalistes stellaris		1.9
Pomadasyidae	1.7	
Pomadasys		1.7
Sciaenidae	1.1	
Pentapodidae	0.8	
Pentapodus caninus		0.8
Serranidae	0.5	
Theraponidae	0.4	
Therapon spp.		0.4

Table 1 (Cont.)

Species	Percentage in the total catch	
Muraenesocidae	0.4	
Muraenesox cinereus		0.4
Stromatidae	0.3	
Pampus spp.		0.3
Heterosomata	0.3	
Lactaridae	0.2	
Lactarius lactarius		0.2
Drepanidae	0.1	
Drepane spp.		0.1
Rachycentridae	0.1	
Rachycentron canadus		0.1
Formionidae	0.1	
Formio niger		0.1
Polynemidae	0.1	

Table 2

Amino acid and chemical composition of several species of trashfish

Amino acid	Mol. weight	Nemipterus	Dorosoma chacunda	Upeneus sp.	Clupea	Saurida sp.
LYS	146.2	6.1	6.9	7.4	7.7	8.2
HIS	155.2	1.7	1.5	2.0	1.7	2.2
NH_3	17.0	0.8	0.8	0.9	1.0	0.9
ARG	174.2	5.6	5.1	5.8	5.5	5.9
HPR	131.1	2.2	0.8	1.4	1.2	1.2
ASP	133.1	8.0	8.0	9.6	8.8	9.3
THR	119.1	3.6	3.5	4.3	4.0	4.6
SER	105.1	3.2	3.1	3.9	3.6	4.5
GLU	147.1	11.5	12.8	14.5	14.1	15.4
PRO	115.1	4.8	3.6	4.6	4.2	4.4
GLY	75.1	8.3	4.9	7.2	6.8	7.2
ALA	89.1	6.6	5.2	7.0	6.5	6.7
CYS	121.1(2)	0.6	0.5	0.7	0.8	0.5
VAL	117.2	3.8	3.9	4.7	4.4	4.9

Table 2 (Cont.)

Amino acid	Mol. weight	Nemipterus	Dorosoma chacunda	Upeneus sp.	Clupea	Saurida sp.
MET	149.2	2.2	2.3	2.7	2.1	3.4
ILE	131.8	3.0	3.3	4.1	3.9	4.4
LEU	131.8	5.8	6.3	7.1	6.7	8.2
TYR	181.2	2.3	2.4	2.7	2.4	3.3
PHE	165.2	3.2	3.2	4.0	3.5	3.7
TRP	204.2					
Glucosamin						
Sum		83.3	78.1	94.6	88.9	98.9
Protein		75.0	77.3	69.8	79.6	78.2
Lipid		4.0	7.1	10.2	3.5	1.9
Ash		22.0	14.4	17.7	17.4	16.4
Sum (dry weight)		101.0	98.8	97.7	100.5	96.5

Data obtained from Prof. Dr. Jan Raa (Norway)

Table 3

Total landing of trashfish in 1977

Species	Tons
Threadfin breams	7 425
Bigeyes	1 474
Croaker/drums	26 340
Slipmouth	36 216
Marine catfish	20 155
Lizard fish	5 270
Goatfish	7 362
Grunters	2 963

Table 4

Utilization of the catch - 1977 (tons)

	Quantity	Percentage
Fresh fish	539 537	46.6
Dried/salted	384 109	33.2
Boiled	69 600	6.0
Fermented		
Terasi	52 472	4.6
Peda	3 798	0.2
Sauce	282	
Smoked	22 384	1.9
Freezing	54 014	4.8
Canning	10 832	0.9
Fish meal	12 138	1.1
Others	8 525	0.7
Total	1 157 691	100.0

3. POSSIBLE SUPPLY OF RAW MATERIAL FOR SILAGE

In order to produce cheap fish silage the price of the fish should be low. Several sources of cheap raw materials are possible.

3.1 Trashfish or by-catch

Trashfish is available along the muddy coastal waters of Indonesia especially in Eastern Indonesian waters, around the Arafura Sea and West Irian. With decreasing catches of shrimp, the percentage of trashfish is becoming higher. This is mostly thrown back to the sea.

3.2 Excess catches of small pelagic fish

One possible source of raw material is the Bali Strait oil sardine. At the height of the season the price of the fresh sardine is so low that the catch is not handled properly. The estimated volume of sardine waste in 1976 was about 8 000 tons at one side of the Bali Strait (East Java).

3.3 Excess of freshwater fish harvest during the dry season

In Sumatera, Kalimantan (Borneo) and Sulawesi (Celebes) during the dry season (of about 3-5 months) so much fish is caught that not enough salt is available to preserve the catch and wastage is very high. The production of freshwater fish in Sumatera in 1977 was 105 671 tons, in Kalimantan 130 352 tons and in Sulawesi 47 869 tons while Java produced 125 119 tons. In the case of Sumatera the amount of waste produced in 1976 and 1977 has been estimated as more than 20 percent.

3.4 Fish offal

In 1977 the total export of frozen shrimp was 29 582 tons of which about 25 percent is waste; while the export of frozen frog legs was 1 980 tons and about 70 percent wasted. The waste of these valuable commodities occurred mostly during fishing, handling and collection activities due to high ambient temperature, simple handling practices and the lack of ice.

4. PROBLEMS

4.1 Freshness of the raw material

Due to the wide-spread locations, especially outside Java, the production of first quality fish silage will be very difficult. Because the price of fresh trashfish is relatively cheap, the costs of collecting and handling will also be as simple and cheap as possible. If the raw material has to be shipped to the hinterland or another island for silage processing, the raw material could be in an advanced stage of spoilage.

4.2 Government policy

In an effort to increase fish consumption and to utilize more trashfish for human consumption, a Government decree was passed in which shrimp fishing companies were requested to at least utilize the more valuable trashfish. As a result of economic considerations, however, only a small part of the catch has been utilized either as fresh fish or mechanically dried fish.

4.3 Availability of acid

To produce acid silage a fairly cheap supply of acid should be available but the acids used to process fish silage are mainly imported. Of sulphuric, hydrochloric, formic and propionic acids, only sulphuric and hydrochloric are produced locally; annual production of sulphuric is about 35 000 tons from two factories while 14 256 tons of hydrochloric acid is produced. The prices of technical grades of acid being sold by chemical suppliers are as follows:

Hydrochloric, 20% - Rp. 10 000/2½ 1[1/]
Formic, 90% - Rp. 15 330/2½ 1
Sulphuric - Rp. 9 690/2½ 1

However, the cost is significantly reduced for larger quantities of commercial grade acids.

4.4 Species composition of trashfish

Typical of the fish in tropical waters the trashfish consists of many species. In order to obtain uniform composition of the fish silage, the mixture of fish species should be considered. Species composition is also influenced by location of fishing grounds and seasonal variation. Thus, it would be preferable to use several species of fish which are caught in larger volumes.

4.5 Practical demonstrations on the use of silage

Further research will certainly make silage processing and utilization better understood. One of the problems which will certainly arise during the practical application of fish silage is how to make the chicken or fish farmer benefit from the simple technology. The farmers will want a clear-cut and simple technique and the beneficial results of the process should be readily apparent.

5. DEVELOPMENT CONSTRAINTS

5.1 Supply of cheap acid

Fish landing places are usually far from consumer centres, and would-be silage consumers. The need for the acid will occur around the landing places or places where there are many chicken farms or other domesticated animals. The acids needed for silage processing should be cheap and readily available.

1/ Rp. 625 = U.S.$ 1

5.2 Preservation and storage of the silage

The production of fish silage is fairly easy to understand and to adopt, but application at the village level should be carefully considered. Since the processing of silage will coincide with the fishing season when fish prices are low, the preservation and the storage of the silage should be carefully planned to avoid mistakes; otherwise the whole effort will be wasted. For example, it is necessary to know how long silage can be stored and when is the right time to use the silage for feeding.

5.3 Fishy taint

It is common knowledge that the use of fish silage or fish meal at high levels will cause fishy taint which will certainly affect the eating quality and the economic value of the eggs or meat. When designing experiments or preparing feeding rations the optimum percentage of fish silage should be used.

5.4 Mechanism for the collection of trashfish

In Eastern Indonesian waters, the problem of how to collect the trashfish from the trawlers with 5-6 weeks fishing operation poses considerable problems. The fish itself is very cheap but collecting the trashfish and bringing it ashore will need careful consideration.

5.5 Transportation and silage containers

For liquid silage a leak-proof container will be needed, and the transportation of the containers will involve technical know-how. If silage production is shore based then the transportation problems are reduced but if the silage is produced at sea, the right type of container should be available and careful disposition of deck space is needed. The choice of the container is made more difficult as the liquid product is preserved by the use of acid.

6. CONCLUSIONS AND RECOMMENDATIONS

6.1 Conclusions

- Due to simple preparation and processing methods, the production of fish silage can be applied at the village level.

- The silage processing does not involve high cost or technology, the utilization of fish through the silage process also requires relatively little energy.

- The process is dependent on the availability of raw materials.

- Since the trashfish is a potential raw material which consists of a mixture of fish species, standard chemical composition and the silage quality should be carefully studied.

6.2 Recommendations

- The production of fish silage onboard vessels needs to be explored.

- Successful application of fish silage at the rural level needs to be explored in order to popularize the new product.

- Extensive extension is needed to promote the use of silage as an important source of protein.

- A techno-economic study on the suitability of fish as raw material for silage and silage production is needed to highlight the advantages and disadvantages of fish silage production.

THE STATUS OF RESEARCH ON FISH SILAGE IN INDONESIA

S. Ilyas
Research Institute of Fishery Technology
Jakarta, Indonesia

I.P. Kompiang
Centre for Animal Research and Development
Bogor, Indonesia

This paper outlines in brief the overall research activities on fish silage in Indonesia. Work on fish silage was started late in 1976, following the feasibility study conducted by J.L. Sumner and discussions between FAO (D. James), TPI, London (J. Disney), LPTP, LPPD and CSIRO. Firstly, we studied the methods of preparation using different acids to determine the amount of acid required. It was found that more acid was required than recommended in European countries to obtain a stable silage.

In 1977, in collaboration with the Centre for Animal Research and Development (CSIRO, Australia) work was started to evaluate the nutritional value of chemical silage when compared to fish meal (prepared from the same materials) by assaying the growth rate of young chicken. The silage was mixed either with cassava, rice bran or rice polish and was then sun-dried. Inclusion of these dry products in the ration for young chickens at low levels (20 percent equal to 4 percent dry silage) gave a similar performance to the fish meal control ration. However, when higher levels of silage were used growth depression was observed (Kompiang, 1978). Since it was not known whether this depression was due to the level of the silage or the filler, it was decided to concentrate on the study of the properties of the silage *per se*, i.e., by absorbing the silage onto corn, which was thought not to contain toxic factors.

The amount of the silage in the rations was also increased to 23-25 percent (dry silage) to accentuate the differences, if any, between fish silage and fish meal. It was found that the nutritional value of the silage was poor when compared to fish meal. During this difficult period, Prof. J. Raa from Tromsø, Norway visited Indonesia as an FAO consultant. This visit greatly stimulated the work and the pace of the research increased considerably. The method of preparing the fish silage was modified including boiling the silage, making silage from boiled fish and extracting the lipid. All of these modifications, which have been reported previously, improved the nutritional value of the silage but never to the level of fish meal (Kompiang *et al.*, 1979a, 1979b). We also supplemented the silage ration with either vitamin B1 or manganese, either singly or in combination, since the birds showed symptoms of slipped tendon and head retraction. However, supplementation had no effect on the nutritional value of the silage (Darwanto, 1979). This result was in conflict to a report from TPI in London that vitamin B1 supplementation improved the nutritional value of the silage.

Work is now continuing in an attempt to identify the problem(s) and to find solutions. Part of the problem is undoubtedly associated with the lipid fraction of the silage and the stability of tryptophan. Current research is attempting to find a way to protect the lipid from oxidation or to find a simple way to extract the lipid. The second alternative is more attractive since the oil itself has a considerable value, especially silage prepared from lemuru (Bali Strait sardine) which contains a very high fat content (up to 15 percent by weight).

Besides evaluating the nutritional value of the silage with chicken, trials are also being carried out on pigs and freshwater fish in collaboration with the Research Institute for Animal Husbandry and the Institute for Inland Fishery. These experiments are directed more toward the application of fish silage rather than the academic aspects as in the chicken experiments. The results will be discussed in detail later.

Recently work is also being carried out on biological fermentation as a way to prepare fish silage. The biological silage has some advantages over chemical silage, at least for Indonesia, where the acids are still imported; they are rather hard to obtain in the village and the price is high. For biological silage, the raw materials are available locally and fermentation is very common in Indonesia for the preservation of surplus food. Some results of the work on biological silage will be presented later.

To summarize, the work on fish silage in Indonesia is attempting to find ways of improving the nutritional value of chemical and biological silage. It is hoped to carry out a pilot commercial project as soon as possible, especially on species which are able to tolerate silage, such as the pig.

REFERENCES

Darwanto, A., 1979 B.Sc. Thesis in Analytical Chemistry Academy, Bogor, Indonesia

Kompiang, I.P., 1978 J.Fish.Technol.

Kompiang, I.P. and R. Arifudin, 1979a Proc.Poultry Science and Industry

Kompiang, I.P., R. Arifudin and J. Raa, 1979b Nutritional value of ensilaged by-catch fish from Indonesian shrimp trawlers. In Proceedings of the Torry Jubilee Conference, Aberdeen, Torry Research Station

PROSPECTS FOR FISH SILAGE PRODUCTION IN MALAYSIA

Wan Johari bin Wan Daud
Malaysian Agricultural Research and Development Institute (MARDI), Malaysia

Abstract

The livestock industry in Malaysia has increased tremendously during the last few years and there is a high demand for feed. Fish meal is becoming an expensive item because most feed producers, for reason of quality, prefer imported fish meal.

The introduction of trawlers has undoubtedly increased the quantity of by-catch. At present the by-catch is either made into fish meal or is dumped at sea. These by-catches and also the offals and trimmings from fish canneries offer good raw materials for making fish silage.

The available raw material and the interest of Government agencies favour the introduction of fish silage in Malaysia.

1. INTRODUCTION

Fish silage, or liquid fish protein as it has been called, has been widely studied in temperate areas culminating in the commercial production of 25 000 t annually in Denmark by 1972 (Disney et al., 1977). Fish silage is used as a protein supplement in the feeding of domestic animals. It is a liquid product preserved with acid and can be made either from whole fish or fish waste.

In the manufacture of silage, fish is ground or mixed with acid which prevents the growth of spoilage bacteria. Either inorganic or organic acids can be used. Alternatively lactic acid production can be encouraged by mixing a carbohydrate source with the ground fish and inducing fermentation. Subsequent liquefaction is brought about by the activity of proteolytic enzymes naturally present in the fish.

Fish silage is being produced commercially in Europe where the main outlet is as pig feed. There is an increasing interest in silage production to take care of small scattered landings where the installation of small fish meal plants would not be economic.

Although the principle of ensilage has been practised in Europe for many decades, there has been little interest in the application of this technique in Malaysia until recently. It is envisaged that the ensiling process will be initiated on the trawlers and continued at the fishing villages after landing. The level of technology employed must therefore be simple, inexpensive and safe. The current interest in Malaysia is very timely, if one takes a close look at the available resources for silage production.

2. AVAILABLE RESOURCES

2.1 By-catch

Trawler fishing operations have expanded considerably since their introduction in 1963 in West Malaysia resulting in an increase in by-catch landings. The increase in by-catch landings in West Malaysia is shown in Table 1. The largest increase in by-catch landings came in 1972 and since then by-catch usually represents the largest component of total catch. In 1977 by-catch composed 31% of total catches or 154 995 t, valued at M.$ 17 316 173 (Anon., 1977).

By-catch may consist of fish, crustaceans and molluscs and it should be borne in mind that every sample of by-catch is bound to be different from the next. The proportion of various fish types differs with season and place of haul. Table 2 gives an example of the proportions of the various fish types found in by-catch.

Table 1

By-catch landing in Malaysia by State, 1970-1977 (t)

	1970	1971	1972	1973	1974	1975	1976	1977
Perlis	60	62	99	2 058	2 783	3 118	2 944	5 980.09
Kedah	4 954	7 814	12 409	16 658	16 790	18 468	17 590	25 094.43
Penang	10 084	11 274	10 761	10 364	7 785	4 851	3 884	9 008.52
Perak	14 634	15 900	14 386	17 001	29 097	22 620	27 930	35 215.12
Selangor	10 792	19 943	26 535	46 025	58 608	50 573	44 638	56 540.71
Negeri Sembilan	-	-	-	-	-	-	-	-
Malacca	247	227	61	86	94	98	88	78.70
Johore	13 998	17 980	13 321	15 573	15 942	13 255	15 113	7 596.19
Kelantan	187	83	227	405	490	667	491	260.79
Trengganu	364	614	729	2 209	6 505	6 981	8 380	9 436.11
Pahang	1 204	1 290	1 880	3 521	8 521	4 556	6 423	5 784.39
Total by-catch landing	56 506	75 166	80 381	113 909	146 567	125 185	127 481	154 995.05
Total marine fish landing	299 005	323 061	311 108	371 230	439 575	373 235	410 968	497 952

Source: Anon., 1970-1977

Table 2

Composition of by-catch as sampled at Kuala Trengganu, Malaysia

Common name	Scientific name	% in batch
Gar fish	*Tylosurus* spp.	8.6
Box fish	Order Tetraodonteidei	14.8
Trevally, horse mackerel	*Caranx carangus*	0.3
Threadfin snapper	*Lutianus nematophorus*	34.4
File fish	Order Rallisteidei	12.5
Catfish eel	*Pletosus canius*	0.3
Catfish	Tachysuridae	0.2
Trigger fish	*Abalistes stellaris*	1.3
Pony fish	Leiognathidae	17.2
Silver biddy	*Gerres abbreviatus*	4.6
Flying fish	Exocoetidae	1.5
Flat fish	Pleuronectiformes	0.9
Soldier fish	*Holocentrus diadema*	0.3
Monocle bream	*Scolopsis vosmen*	1.5
Lizard fish	*Saurus myops*	0.9

The by-catch has been sorted and the more important species have been analysed for moisture, protein, fat and ash. The results are shown in Table 3.

Table 3

Proximate analysis of by-catch species

Common name	Scientific name	% Moisture	% Protein	% Fat	% Ash
Threadfin snapper	*Lutianus nematophorus*	69.44	21.51	3.58	4.90
Monocle bream	*Scolopsis vosmen*	66.55	20.2	3.94	7.55
Pony fish	Leiognathidae	69.20	22.41	1.18	5.62
Flat fish	Pleuronectiformes	70.07	21.38	2.33	6.00
Silver biddy	*Gerres abbreviatus*	70.21	21.82	1.88	5.19
Gar fish	*Tylosurus* spp.	73.76	15.86	1.32	7.33
Flying fish	Exocoetidae	66.56	20.64	2.43	8.46
Soldier fish	*Holocentrus diadema*	69.01	19.56	3.46	6.39
Lizard fish	*Saurus myops*	68.75	20.5	0.69	7.38
Trigger fish	*Abalistes stellaris*	69.72	21.0	1.47	6.79

The analytical figures showed that the protein content of the by-catch is high. This protein should not be wasted.

2.2 Offals and trimmings

(a) Canneries

Fish waste from processing industries such as tuna canning can also supplement by-catch as raw materials for the production of fish silage. Currently there are few fish canning plants, but with a projected catch of 900 000 t by 1995 (Table 4), a growth of cannery plants may occur. Offals and trimmings from such plants could become an important source of raw material for silage.

Table 4

Projected catch/by-catch from 1980-1995 (t)

Year	Projected catch	Projected by-catch
1980	550 000	165 000
1985	680 000	204 000
1990	790 000	237 000
1995	900 000	270 000

By 1995 it has been predicted that 93 new factories will be built with an estimated investment value of approximately M.$ 465 million[1/] (Abdullah et al., 1978). This projection, however, is subject to the scale of development that could be undertaken by the Fisheries Development Board as well as the Fisheries Department of the Ministry of Agriculture and the development of technology by agencies like MARDI.

Wastes from the canning factories can be in two forms. First, heads, tails and guts amount to 35% waste. The second type of fish waste is in the form of red meat or 'blood meat' (if not canned for pet food) which normally forms 25% of waste (Idrus et al., 1978).

(b) Small fish processors

Table 5 lists the products made from the marine catches in 1977. In the production of these products the guts, heads, tails and bones are normally wasted. The products like salted dried fish, fish crackers, prawn crackers, smoke tuna, peeled prawns all produce offals and trimmings ideal for fish silage.

2.3 By-catch from shrimp fishing which is discarded at sea

Disney (1979) suggested that in 1977, of the total marine catch (approximately 500 000 t), about 15% would be completely wasted or poorly utilised. This wastage would arise principally from shrimp by-catch, approximately 60% of which would be dumped at sea and 40% would be brought ashore. Unfortunately, these figures cannot be verified due to the difficulty of estimating the quantity of fish dumped at sea.

By-catch is dumped first because it realizes a very low price per unit weight and second the fisherman prefers to keep the space available in case of possible bumper catches later in the trip. For example, in Malaysia boats generally make day trips but if catches are poor

---

1/ M.$ 2.17 = U.S.$ 1

they may be extended to 36 hours. In these circumstances the by-catch may be dumped if the boat is fishing for shrimps. However, if it can be demonstrated to the fishermen that the by-catch can be converted profitably into fish silage then the wastage of by-catch could be avoided.

Table 5

Processed marine fish products 1977 (t)

Type of Product	
Dried salted fish	6 793.10
Dried anchovies	6 580.64
Boiled fish	720.93
Manure fish	23 596.30
Fish meal	11 927.40
Dried prawn	3 057.59
Prawn dust	10.70
Prawn paste	111.67
Shrimp paste	1 512.58
Fish crackers	1 578.76
Prawn crackers	17.90
Prawn sauce	3.44
Smoke tuna	110.16
Cuttle fish	19.29
Cooked frozen prawn	278.58
Fish sauce	4.78
Jelly fish	592.13
Dried cockles	6.04
	56 921.99

Source: Anon., 1977

3. METHODS OF INTRODUCTION

In order to achieve the most effective introduction of a new technology, the relevant Government agencies should work closely together. In Malaysia, the relevant agencies are MAJUIKAN (the Fish Development Authority), Fisheries Department, MARDI, MAJUTERNAK (the Livestock Development Authority) and the Veterinary Department.

MAJUIKAN is expanding its trawler fleet which will produce large quantities of by-catch. The fishermen's association, that is the association formed to look after the welfare of the fishermen, comes under the administration of MAJUIKAN. They could therefore spearhead the introduction of fish silage to the fishery.

MARDI with its Agricultural Product Utilization (APU) and Livestock Production (LP) Divisions can concentrate on the formulation of the product and carry out feeding trials. At present, the APU division of MARDI is concentrating mainly on the utilization of by-catch for silage production. Cassava flour is being employed as the carbohydrate base (Yeoh and Merican, 1978).

The Livestock Production (LP) division of MARDI will be able to test the formulated silage diets on pigs and poultry.

MAJUTERNAK (Livestock Development Authority) is the government agency involved in the rearing and marketing of livestock. They are in a position to utilize the successful technology which has been developed by the research agencies. With the existing government agencies one can forsee that silage production is not a difficult thing to introduce in Malaysia.

4. CONCLUSIONS

The availability of resources for silage production favours the introduction of this product in Malaysia. Disney (1979) mentioned, there are four factors favouring the commercial production of acid silage in Malaysia:

(a) There are large quantities of waste fish in the vicinity of commercial piggeries;

(b) In many cases, fish meal production is costly; a low quality product results and losses may be considerable;

(c) With the large use of formic acid in the rubber industries, formic acid is relatively cheap;

(d) Animal feeds are costly as many of the materials are imported.

For developing countries like Malaysia, fish silage is advantageous because it can be introduced on both the cottage industry and the commercial scales. The process is not complicated and minimal amount of training is required for the processors (Sumner, 1976).

At the present rate of expansion of livestock industry in Malaysia especially the pig and poultry industries there is a good prospect for silage production in Malaysia.

5. REFERENCES

Abdullah, M.I. and A.Z. Idrus, The fish processing industry in Peninsular Malaysia.
1978 Proc.IPFC, 18(3):45-60

Anon., Annual Fisheries Statistics 1977. Ministry of Agriculture, Malaysia
1977

Disney, J.G., Prospects for fish silage in Malaysia, Sri Lanka, Bangladesh and the P
1979 Philippines, IPFC Occas.Pap., (1979/1):51 p.

Disney, J.G., I.N. Tatterson and J. Olley, Recent developments in fish silage. In
1977 Proceedings of the Conference on the handling, processing and marketing of tropical fish. London, Tropical Products Institute, pp. 231-40

Idrus, A.Z., W. Rahimah and J. Abdullah, Fish meal processing; status and prospects in
1978 Malaysia. Proc.IPFC, 18(3):61-72

Sumner, J., A feasibility study for fish silage production in the Indo-Pacific Fisheries
1976 Council region. Paper presented to the Second Meeting. IPFC Working Party on Fish Technology and Marketing, Colombo, Sri Lanka, 25-26 October 1976. Rome, FAO, IPFC:FT/76/4:40 p.

Yeoh, Q.L. and Z. Merican, Processing of non-commercial and low-cost fish in Malaysia.
1978 Proc.IPFC, 18(3):572-80

THE STATUS OF RESEARCH ON FISH SILAGE IN MALAYSIA

Yeoh Quee Lan
MARDI, Malaysia

Abstract

The quantity of by-catch landed in Peninsular Malaysia forms a substantial proportion of the total marine landings of the country. Some of this by-catch is used to produce fish meal and fish manure but approximately 30 percent is not utilised, resulting in annual losses of millions of Malaysian dollars. On the other hand, substantial amounts of protein supplements for the animal feed industry are imported. Production of fish silage may provide a solution to both problems.

Research on fish silage production in Malaysia has concentrated on biological fish silage. The results to date have been promising and animal feeding trials have shown that the product is acceptable. Preliminary trials on acid fish silage have also been initiated. This paper reports the results obtained to date.

1. INTRODUCTION

Fish is one of the most important sources of protein in South East Asia. Many preservation methods have been developed, such as drying, salting, fermentation, freezing and canning. In spite of these practices, considerable wastage occurs through spoilage resulting from inadequate storage, handling and processing systems, and the discarding of the by-catch, especially in the prawn trawling industry.

Research has therefore been initiated to develop alternative methods for preserving this protein resource. Ideally, the methods used should be inexpensive and easily adapted for use at the village level. Production of fish silage appears to fulfil these requirements and can be used either as a supplement or precursor to fish meal production.

2. AVAILABILITY OF FISH

The total amount of fish landed in Peninsular Malaysia is substantial both in monetary value and quantity. In 1977, the total marine landing was 497 952 t which retailed at a value of M.$ 970 347 181 1/. This includes manure fish which was valued at M.$ 24 971 373 (Annual Fisheries Statistics, 1977). Table 1 shows the percentage by-catch in Malaysia for the years 1970-1977.

Thus it can be seen that the by-catch makes up a large proportion of the total marine fish landing. If the figures for trawl landings only are considered (Table 2), it is obvious that most of the by-catch is a result of the trawl fishery.

Most of the by-catch is used for the production of fish meal and manure fish. A small proportion is used for making various preserved products such as salted fish (Table 3). However, a large proportion is not utilized and it is this surplus which could be converted to fish silage.

3. SILAGE PRODUCTION

Silage production could have many advantages for developing countries as it makes use of fish which is currently wasted and does not required expensive equipment or sophisticated knowledge. The scale of operation can also be easily varied depending on the supply of fish

---

1/ M.$ 2.17 = U.S.$ 1

available. It may even be initiated on board ship so that spoilage of the trash fish can be prevented. Very often the trash or by-catch fish is kept un-iced on the decks of trawlers because of the inhibitive costs of ice. The fish therefore arrives at the fish meal factories in a poor condition. Thus, if the preservation process was initiated on board the ship there would be no loss of valuable protein due to spoilage and the resultant fish silage could be used directly as an animal feed or as a precursor for the manufacture of fish meal.

The principle of ensilage is the preservation of the material by preventing microbial degradation through the addition of chemicals or biological derivatives. No information is available on research being carried out in other institutes in Malaysia but at MARDI the work has concentrated on the production of biological fish silage. Work on acid fish silage has been initiated only recently.

Table 1

By-catch landings in Peninsular Malaysia - 1970-1977 (t)

Year	Total Marine Landing	By-catch Landing	Percentage By-catch
1970	299 005	56 506	18
1971	323 061	75 166	23
1972	311 108	80 381	25
1973	371 230	113 909	30
1974	439 575	146 567	33
1975	373 235	125 185	33
1976	410 968	127 481	31
1977	497 952	154 996	31

Source: Adapted from Annual Fisheries Statistics, 1970-1977

Table 2

Trawl landings in Peninsular Malaysia - 1972-1977 (t)

Year	Total Landings	Fish	Prawns	By-catch	Percentage By-catch
1972	111 722	27 956	25 740	58 026	51
1973	153 713	37 805	31 135	84 773	55
1974	195 594	46 787	31 449	117 358	60
1975	182 888	49 874	29 328	103 686	56
1976	220 679	72 917	36 200	111 562	50
1977	261 125	85 321	42 197	133 607	51

Source: Adapted from Annual Fisheries Statistics 1976, 1977

Table 3

Utilization of by-catch for the production of fish meal and fish manure 1970-1977 (t)

Year	Total by-catch	Fish meal	Fish manure	Total utilized	Percentage underutilized
1970	56 506	33 686	20 890	54 576	3
1971	75 166	26 088	24 443	50 531	32
1972	80 381	21 320	30 541	51 861	35
1973	113 909	25 226	28 570	72 835	36
1974	146 567	56 431	38 402	94 833	35
1975	125 185	48 810	25 862	74 672	40
1976	127 481	30 829	23 296	54 125	42
1977	154 996	47 753	39 327	87 080	43

Source: Annual Fisheries Statistics, 1970-1977

3. BIOLOGICAL FISH SILAGE

Silage was prepared by the following method. A known quantity of whole fish was rinsed, drained and minced using a large bowl chopper/grinder for 2-3 min. Ground cassava chips, ragi and starter culture (and salt if required by the formulation) are added and the mixture is minced for a further 2 min. The mixture is then packed into airtight containers which are kept at room temperature (28-30°C) for the fermentation process. Ragi is a starter used locally for some fermented products and contains strains of amylolytic fungi (Stanton and Yeoh, 1976) which are necessary to break down the carbohydrates to sugars. The lactic acid bacteria starter culture is prepared from sauerkraut equivalent fermentations and contains a mixture of *L. mesentereides*, *P. cereviseae*, *L. brevis* and *L. plantarum* with a count of 1.0 to 3.0 x 10^8 organisms/ml and a titratable acidity of 1.0 to 1.2 percent as 'lactic' acid.

The formulations which have been found to produce good silage are as follows:

(a) fish carbohydrate in the ratio 1:1; 5% ragi and 2% starter culture by weight are then added

(b) 80% fish, 15% carbohydrate, 5% ragi; 2% starter culture and 4% salt by weight are then added.

In terms of practical application and economics the second formula is more suitable for large-scale production as less carbohydrate is required. The resulting silage may be fed direct to animals or dehydrated and stored until required.

The changes which occur during fermentation, such as pH, percent 'lactic' acid and microbial flora, have been monitored (Yeoh, 1979). Good quality silages were found to have the following characteristics:

1. A rapid drop in Ph from about 6.0 or 6.5 to below pH 5.0. The more successful the fermentation, the more rapid the drop and the lower the final pH value

2. A high 'lactic' acid content. The level usually increases sharply during the first few days, and remains fairly constant for the rest of the fermentation

3. The ammoniacal nitrogen content is low

4. A low anaerobic spore former and coliform count

5. No pathogens such as <u>Salmonella</u> spp. or <u>Staphylococcus</u> spp.

6. An acceptable fishy smell

7. The volume of gas generated during fermentation is relatively small

8. Remains stable for more than six months in the wet form and for more than one year in the dehydrated form.

Feeding trials carried out with mice and chickens (Yeoh, 1979) showed that the fish silage was suitable for use either as a complete or partial replacement in the diet of young animals. Acceptability by the trial animals was good and no ill-effects were observed. A fish silage dehydrated to approximately 10% moisture has a yield of about 50% of the original wet weight.

The dehydrated fish silage was found to have the following composition:

Crude protein	-	34 - 36%
'Lactic' acid	-	5 - 7%
Ammoniacal nitrogen	-	0.1 - 0.3%
Moisture	-	4 - 6%
Sodium chloride	-	7.5 - 8.5%
pH	-	4.7 - 4.8

By comparison the crude protein content of fish meal produced in Malaysia ranges from 40-50% (Sumner, 1976). Thus, the crude protein content of biological fish silage is not very much less than fish meal.

4. ACID FISH SILAGE

Acid fish silage was prepared by mincing whole fish and mixing it with various proportions of acid. Trials were carried out with organic acids only and with mixtures of mineral and organic acids. The mixtures were kept in acid-resistant containers at room temperature and covered to keep out insects. The silages liquefied after 2 or 3 days into a thick slurry. If allowed to stand, a layer of clear liquid formed at the top of the slurry.

Gildberg and Raa (1977) advocated the use of 0.75% (v/w) propionic acid and 0.75% (v/w) formic acid to produce a stable product with a final pH of 4.5. Trials carried out on trash fish obtained locally showed that these concentrations of acid were insufficient to preserve the silage.

The addition of 2% (v/w) of 98% formic acid or 3% (v/w) of 85% formic acid were also used to preserve the fish. The pH dropped from about 6.5 to 3.8 and 4.4 respectively after addition of the acid and remained relatively stable after that. The addition of propionic acid alone was not as effective and 3% (v/w) of 98% propionic acid only lowered the pH to 4.9 rising to 5.4 after 2 weeks. Trials using a combination of acids, that is sulphuric acid to lower the pH to 4.0, followed by an addition of 0.5% (v/w) formic or 0.5% (v/w) propionic acid or both organic acids have also so far failed to produce stable products. Further trials are being carried out to determine the minimum levels of these acids needed to preserve the silage.

Microbial counts carried out on the acid fish silages showed a drop in the total viable counts and coliforms after addition of the acid. However, if insufficient acid was added, the counts rose again as spoilage developed. On the other hand, total viable counts for the stable products, viz., with 2-3% formic acid, remained constant in the region of $1.3x10^4$ to $5.7x10^4$ organisms/g.

5. CONCLUSIONS

Research carried out to date indicates that biological fish silage can be successfully produced in developing countries like Malaysia. Initial experiments on acid fish silage have shown that the fish can be preserved satisfactorily with 2% (v/w) of 98% formic acid or 3% (v/w) of 85% formic acid. Use of other acids or a combination of organic and mineral acids is still being investigated.

Fish silage production has a number of advantages for developing country situations and shows promise as a supplement or precursor to fish meal.

6. ACKNOWLEDGEMENTS

The author wishes to acknowledge the support and interest of the Head of the APU Division and the kind permission of the Director General of MARDI.

7. REFERENCES

Gildberg, A. and J. Raa, 1977 Properties of a propionic acid/formic acid preserved silage of cod viscera. J.Sci.Food Agric., 28:647-53

Stanton, W.R. and Yeoh Quee Lan, 1977 Low salt fermentation method for conserving trash fish under S.E. Asian conditions. In Proceedings of the Conference on the handling, processing and marketing of tropical fish. London, Tropical Products Institute, pp. 277

Sumner, J., 1976 A feasibility study for fish silage production in the Indo-Pacific Fisheries Council region. Paper presented to the Second Meeting. IPFC Working Party on Fish Technology and Marketing, Colombo, Sri Lanka, 25-26 October 1976. Rome, FAO, IPFC:FT/76/4:40 p. (mimeo)

Yeoh, Q.L., 1979 Fermentation methods for the preservation of fish and fish trash. M.Sc. Thesis: University of Malaya

THE STATUS OF FISH SILAGE RESEARCH IN THE PHILIPPINES

Susan Canonizado
Bureau of Fisheries and Aquatic Resources
Manila, Philippines

1. INTRODUCTION

The optimum utilization of fish is attained when all parts are converted into by-products of commercial value or no wastage is allowed in the fishing industry.

In the Philippines, the main effort is concentrated on the preservation of fish for human food. However, there are situations when this is not possible. Seasonal variations in catch, transport difficulties, lack of ice and cold storage facilities, inadequate processing facilities, etc., all occur such that considerable amounts of fish are wasted, discarded at sea at the moment of catch or left unutilised. These underutilized fish may be converted into fish sauce, fish paste and fish meal; the majority are channelled into the fermenting industries (fish paste and fish sauce) which results in the surplus production being exported. Conversely, fish meal is imported which represents an appreciable and constant drain on the foreign currency balance.

The Philippine fish meal industry has considerable potential for development due to the increasing demand for feed ingredients. According to the Bureau of Animal Industry, the local fish meal industry was only capable of supplying about 20 percent of the requirement for fish meal in 1978. The local pig and poultry industries alone require so much fish meal that 18 133 t of fish meal were imported in 1978.

However, because of the high capital investment involved in the manufacture of fish meal, there is a need for the development of a protein supplement for animal feeds that would require less sophisticated equipment and consequently a lower capital cost.

The acid ensilage of fish as an alternative to conventional fish meal has been suggested by many workers. The basic methodology was developed more than 50 years ago and substantial quantities are produced in countries such as Denmark and Poland (Disney et al., 1978). In recent years, attention has been concentrated in the development of fish silage as a means of utilizing fish wastes in situations where fish meal production is not available.

Sumner (1977) in his study on fish silage in the Indo-Pacific region concluded the following advantages of fish silage production: the operation is simple and does not require intensive training; capital investment is low; it can be produced aboard the fishing vessel so that no chilling storage of fish is required prior to ensilage; the process is temperature dependent and occurs at an extremely rapid rate at tropical temperatures and silage production can be produced by large and small-scale industry.

The development of fish silage in the Philippines will not compete with the fish fermentation and fish meal industries but would rather assist in the complete utilization of available resources and thereby result in a lower cost for animal feeds.

2. AVAILABILITY OF RAW MATERIALS

Total fish production in the Philippines in 1977 was about 1.5 M t of which 39.7 percent was caught by trawl fishing (Tapiador, 1979). Trawl fishing results in the catching of small fish and other organisms collectively known as by-catch. This is estimated to reach as high as 40 percent of the total fish catch.

The by-catch is usually dumped overboard at sea as the fishermen are not aware of its considerable value and also because they prefer to keep space available for possible catches later during the trip.

There are other sources of raw materials such as the 30-40 percent fish wastes in canning plants, the 10 percent rejects in drying establishments and the 2-3 percent fish wastes in smoking plants.

3. THE FISH SILAGE PROCESS

Fish silage is a protein supplement added to animal feeds. It is a liquid product that is preserved with the use of acids. The acid environment activates the breaking down of protein into small soluble units converting the fish into a semi-liquid form, and produces an unfavourable condition for bacterial growth. Proteolytic enzymes, naturally present in the fish, contribute to the hydrolysis by breaking down the peptides not fully hydrolyzed by the acid and subsequently, liquefaction of the product is attained.

Acids such as sulphuric, hydrochloric or formic acid may be used alone or in combination. Formic acid has been said to be a better choice than mineral acids because preservation is achieved at a slightly higher pH, has some bacteriostatic action and the silage need not be neutralized before adding it to the feed. However, formic acid is more expensive.

Another process of making fish silage is the development of lactic acid from the bacterial fermentation of sugars. In a study by Arifuddin _et al._ as cited by Tapiador (1979), lactic acid fermentation starts instantly when crushed terrestrial snails are mixed with corn or cassava resulting in a pH drop to 3.8 within 24 hours at 30°C.

4. RESEARCH TO DATE ON FISH SILAGE

Santos E. _et al._ (1977) have determined the concentration of sulphuric acid that should be added to milkfish offal in order to produce a good quality silage within the shortest span of time. The addition of either 25 percent or 30 percent of sulphuric acid in an amount equal to 15 percent of the weight of the raw materials was recommended. The final product has a pH of 2.1 after four weeks. A comparison of the amino acid composition of fish meal and fish silage prepared from milkfish offal is shown in Table 1 below.

Table 1

Amino acid composition of fish meal and silage prepared from milkfish offal (Santos _et al._, 1977)

Amino acid	Amino acid content/mg/g protein		Silage
	Dry reduction	Wet reduction	
Lysine	67.24	55.80	43.91
Histidine	27.75	19.59	21.83
Arginine	59.02	46.09	44.17
Aspatic acid	91.08	74.71	60.00
Threonine	41.55	35.52	26.50
Serine	45.12	36.93	29.08
Glutamic acid	126.16	111.52	86.42
Proline	57.06	44.90	53.08
Glycine	95.10	76.10	99.08
Alanine	67.10	57.93	55.83
Cystine	-	Tr	-
Valine	51.78	39.43	29.83
Methionine	0.85	10.41	0.92
Isoleucine	39.51	34.67	22.17
Leucine	69.26	58.43	38.92
Tyrosine	16.45	17.19	8.75
Phenylalanine	37.76	32.12	26.33
Tryptophan	-	-	-

In an earlier study Flores, 1973 found that the net protein utilisation value for fish silage is higher than that for fish meal probably because fish silage undergoes a digestion process which does not occur in the fish meal process (see Table 2). This study also found that the preparation of fish silage is much simpler in approach than fish meal and does not involve heating. The disadvantage however is the need for a non-corrosive container and the great space needed during storage and transport.

Table 2

Comparison between the results of pepsin digestibility of fish meal and fish silage (Flores, 1973)

Samples	Pepsin digestibility % protein	Solubility % protein	% NPU of total protein
	Average	Average	Average
Fish silage	1.1	18.3	36.9
Fish meal	1.8	15.8	33.3

5. PRIORITY RESEARCH NEEDS

Fish silage technology in some countries has already been established. However, in the Philippines there is still a need for considerable research. Work is also required on the use of controlled fermentation (lactic acid) instead of the use of acid.

In the Philippines fish meal production is an acceptable product for use in animal feeds. Since fish silage is still new to the industry the technology involved needs to be publicised. Pilot/demonstration trials of fish silage production should be carried out. Feeding trials, as well as cost benefit analysis on the use of fish silage, are priority areas which must be looked into for fish silage technology in the Philippines to be developed.

6. REFERENCES

Disney, J.G., W.H. Parr and D.J. Morgan, Fish silage: preparation, utilisation and pros-
1978 pects for development. Proc.IPFC, 18(3):543-53

Disney, J.G. et al., Development of a fish silage/carbohydrate animal feed for use in the
1978 tropics. Trop.Sci., 20(2):129-44

Flores, F.A., The nutritive value of fish silage compared to fish meal. Thesis submitted
1973 to the Department of Fisheries Technology of the College of Fisheries, University of the Philippines

Santos, L.M., F. Magno-Orejana and M.N. Bautista, Utilisation of by-product of milkfish
1977 processing. Preparation of fish meal and silage. Fish.Res.J.Philipp., 2(1): 56-69

Sumner, J., Fish silage production in the Indo-Pacific region: A feasibility study. IPFC
1977 FT/76/4 (mimeo)

Tapiador, D., Needs and requirements of fisheries Extension services on fish silage develop-
1979 ment. Paper presented at the ASEAN Seminar/Workshop on fisheries extension Metro Manila, the Philippines, 20 February 1979

PROSPECTS FOR THE PRODUCTION AND UTILIZATION OF FISH SILAGE IN THAILAND

Bung-orn Saisithi and Pong Pen Rattagool
Fishery Technological Development Division
Bangkok, Thailand

1. INTRODUCTION

At the suggestion of the Secretary of the Indo-Pacific Fishery Commission (IPFC), this study was undertaken as a continuation of the sub-regional study previously undertaken by Mr. J. Disney of the Tropical Products Institute, London, entitled "Prospects for Fish Silage in Malaysia, Sri Lanka, Bangladesh and the Philippines". IPFC Occasional Paper 1979/1, January 1979.

As with the earlier study, the United Nations Environment Programme (UNEP) Regional Office for Asia and the Pacific and the IPFC Secretariat/FAO Regional Office for Asia and the Far East provided support services for the preparation of this paper.

2. PRESENT SITUATION

Thailand ranks among the top ten fishing nations of the world with an annual production of 2.26 million tonnes in 1978. About 40 percent of this total were trashfish or "by-catch". The total production of fish has increased steadily from 1974 to 1977 but the quantities of trashfish fluctuated during the same period due to the oil crisis (Table 1). The production of fish meal has grown in direct proportion with the available trash fish except that trash-fish landings decreased slightly in 1976.

Table 1

Total production, trashfish and fish meal production in Thailand (1974 to 1977)

('000 t)

	1974	1975	1976	1977
Total production	1 510	1 515	1 699	2 190
Trashfish	690	635	621	837
Fish meal	94	92	113	138

No records are available on the utilisation of trashfish in 1976; however, statistical data showed that in 1977 fish meal was produced from 572 000 tons of trashfish and 46 000 tons of other fish and shellfish which presumably consisted primarily of small pelagic fish. The remainder of the trashfish, about 260 000 tons was used for duck or fish-feeding fertiliser; a further 5 000 tons was converted into fish sauce.

The fish meal industry in Thailand produces a medium quality meal but the quality is not consistent. This is due to the many small plants with a capacity of less than 5 tons. There are 73 plants in the country and more than 50 percent have capacities of less than 5 tons. The protein content of the fish meal produced varies from 50 to 65 percent. The export of fish meal increased drastically from 27 000 tons in 1975 to 102 000 tons in 1978 valued at more than Baht 700 million (U.S.$ 35 million). Not so many years ago Thailand used to import small amounts of fish meal from Angola and South Africa.

3. FISH WASTE

The fish meal industry in Thailand has virtually reached saturation point, and it is not possible to consider the raw material potentially available for silage production without considering the fish meal industry. It has been reported that trash fish from small- and medium-sized vessels, which normally take 7-10 days each trip, are not being discarded due to a shortage of raw material. In 1976 it was estimated that about 400 000 tons of trashfish per year are dumped overboard at sea. It is very difficult to arrive at any firm data on the quantities of fish which are discarded. However, for the bigger vessels that go on long voyages, some of the low-value species of fish may be discarded due to limited freezers and storage facilities and ice.

There are a number of other sources of fish waste which could be used as raw material for fish silage production. For example, the waste from freezing plants includes shrimp heads, shells and fish offal. Inquiries carried out at 30 freezing plants indicated that the waste is about 15 000 tons/year. This figure does not include squid and cuttlefish offals which are mostly removed at the landing places. An estimated figure from frozen squid and cuttlefish exports shows that squid and cuttlefish offal could be around 8 000 tons/year. Considering that 96 500 tons of raw material is converted into frozen products and assuming that 25 percent is waste, then some 24 000 tons per annum is wasted. If frozen products, dried shrimp and canned fish are considered to be in the same group, the total raw material is about 132 596 tons which will give an estimate total waste of about 33 000 tons/year.

The second category of wastes arises from dried, salted, steamed-smoked and fermented fish products. The total raw material in these products is 270 000 tons which will give an estimated waste of about 27 000 tons/year. The third item of waste arises from fish ball production. A recent marketing survey estimated that about 16 500 tons/year of whole fish are required for fish ball production. The waste from this item could therefore be around 10 000 tons/year.

Although it is extremely difficult to obtain accurate figures for the level of waste in the Thailand fishing industry the estimates suggest the total amount could be in the region of 70 000 tons/year (33 000 + 27 000 + 10 000).

4. OTHER PROTEIN WASTE

There is relatively little waste from poultry and other animals due to the fact that the majority of the offal and blood is eaten. The poultry industry in Thailand is well developed but most parts of the chicken are utilized either directly or indirectly for human consumption. Some parts, such as feathers, are converted to feather meal for export but it was not possible to obtain information on the quantities involved.

For other animals, such as cattle, there is also little waste. Even the bones are ground and used for animal feeding and some bones are used for glue production.

5. THE LIVESTOCK INDUSTRY

The livestock industry in Thailand has changed rapidly in recent years, especially the poultry industry. In the past, sixty percent of poultry (chicken and duck) production was represented by small-scale production at the village level and no concentrates were given to the birds at all although paddy was occasionally fed as a supplement. In recent years, about 70 percent of poultry is raised on compound feed. Most feed companies in Thailand advance credit to commercial poultry farms in the form of feeds, medicines, stocks or even cash. These companies stipulate that their customers must sell the finished products to companies in return. As a result of this arrangement, and a well developed industry, Thailand has considerably increased its exports of chicken as shown in Table 2.

Table 2

Quantity and value of frozen chicken and swine exported (1976-1978)

Year	Chicken		Swine	
	Kg	Million Bahts	Kg	Million Bahts
1976	1 234 283	49.37	33 110	1.79
1977	3 859 513	146.80	95 796	5.17
1978	9 907 214	396.29	1 171 055	63.24

Source: Department of Livestock Development

From Table 2 it can be seen that exports of chicken and pork increased sharply during 1976-1978. The potential for increased chicken export is great due to the high demand in foreign countries. By contrast the export of animals such as cattle and buffalo decreased (Table 3).

Table 3

Quantity and value of livestock and products exported (1972-1974)

	1972		1973		1974	
	Number	Baht	Number	Baht	Number	Baht
Breeding cattle (head)	3	9 000	230	439 700	606	1 919 562
Breeding buffalo (head)	120	355 680	86	276 576	50	161 400
Cattle for slaughter (head)	25 806	39 241 971	25 847	57 963 299	17 283	55 875 064
Buffalo for slaughter (head)	25 055	75 847 097	20 376	79 939 495	11 972	62 725 704
Cattle and buffalo hides (tons)	10 757	134 688 059	6 469	177 111 816	2 864	106 677 073
Swine (head)	1 169	846 366	10 846	8 320 926	2 520	7 269 194
Poultry (bird)	1 120 726	2 096 516	1 705 620	5 273 161	2 335 383	8 885 556

Source: Department of Customs, Bangkok

Changes in the livestock population are shown in Table 4. The number of animals has increased steadily, except that the number of ducks decreased in 1978. Therefore, the demand for animal feeds has also increased.

Table 4

Livestock population 1963-1978 (1 000 head)

	1963	1968	1973	1978
Buffalo	5 147	5 550	5 942	6 562
Cattle	3 624	4 290	4 335	4 705
Swine	3 284	4 503	4 460	4 943
Duck	6 547	6 867	11 078	9 103
Chicken	43 103	53 661	61 816	65 324

Source: Department of Livestock Development, Division of Agricultural Economics

6. ANIMAL FEED REQUIREMENTS

Approximately 70 percent of poultry and 20 percent of pigs in the country are using feedstuffs containing animal protein which comes mainly from fish meal. Other ingredients such as soybean, rice bran, cassava and maize are available locally. It is very difficult to estimate the amount of fish meal used for animal feeding. The quantities of animal feeds produced from 1973 to 1977 are shown in Table 5. Assuming that these animal feeds contain 10 percent fish meal then in 1977 about 72 550 tons of fish meal were used to produce animal feed. The Government allows the animal feed plants to export 10 percent of their production. Therefore, according to these estimates, 65 000 to 70 000 tons of fish meal were used for domestic consumption.

Table 5

Animal feed production (1973-1977)

Year	Quantity of animal feed (tons)
1973	241 986
1974	284 747
1975	486 533
1976	666 353
1977	725 508

Source: Department of Business Economics, Ministry of Commerce

Fish meal for domestic use was estimated at 56 500 tons in 1973 by the Department of Business Economics (DBE) while the Department of Fisheries (DF) estimated that domestic usage was 67 450 tons in the same year. The Department of Business Economics also estimates that livestock production increases to an annual rate of 5 percent. The following tonnage of fish meal for domestic use (Table 6) can thus be calculated:

Table 6

Estimated quantities of fish meal for domestic use (1973-1977)

Year	DBE estimated	DF estimated
1973	56 500	67 500
1974	59 400	70 800
1975	62 400	74 300
1976	65 500	78 000
1977	68 800	81 900

Statistical data from the Department of Fisheries showed that in 1977 the production of fish meal was 138 304 tons, while the amount of fish meal exported was 75 618 tons. Therefore, fish meal for domestic use was about 62 600 tons, thus the Department of Business Economics estimate was the more accurate.

The Livestock Department has also estimated that 5 million tons of animal feed are needed for maximum use in the livestock industry. However, only one third of the animals at present are raised by using animal feeds. The level of fish meal used in animal feed can range from 0 to 10 percent. Assuming a usage rate of 6 and 8 percent, the amount of fish meal required would be 78 000 and 104 000 tons respectively for 1.3 million tons of animal feeds. The figure of 78 000 tons should be more realistic than 104 000 tons. Using an annual rate of increase of 5 percent fish meal for domestic use in 1978 should be around 80 000 tons. Therefore, the production of fish meal in 1978 would be 180 000 tons (100 000 tons for export, as reported by the Department of Customs). The raw material or trashfish used for the production of this amount would be 810 000 tons which is equivalent to the maximum trashfish production (conversion ratio of fish to fish meal is 4:5).

7. FISH FEED REQUIREMENTS

Fish farming in Thailand is considered as an industry. The requirement for fish feed will become more pressing in the future because the marine catch is becoming limited. Trashfish are already being used as feed for farmed catfish and snakehead whose production in 1976 was 6 116 and 4 887 tons respectively. Feed conversion for catfish is 1 kg of catfish from 5.29 kg of trashfish while 1 kg of snakehead requires as much as 9 kg of trashfish. Therefore, the total requirement of trashfish in 1976 for culture of freshwater fish was about 70 000 tons. This figure does not include trashfish used for seabass and shrimp culture on which no data are available.

8. SUMMARY OF THE EXPERIMENTAL WORK CARRIED OUT IN THAILAND

(1) Fish silage has been prepared by acidifying trashfish, fish offal and other fish surplus. Mixtures of sulphuric acid and formic acid can be used and in some cases propionic acid was required to protect against fungi and moulds. The use of sulphuric acid is intended to reduce the cost of production since it is made locally. The proportion of sulphuric acid to formic acid required depends upon the quality of the fish.

(2) Three to five days are sufficient to liquefy the fish. The liquid product can be fed directly to pigs or produce a mixed feed for poultry. The liquid product can be kept for at least six months, However, it is more convenient to dry the product since Thailand is in a tropical zone. Drying should be carried out after five to ten days.

(3) Feeding trials have been carried out to compare fish silage and fish meal with fish, pigs and chickens. The growth of chicken was comparable for both fish meal and fish silage. No palatability problems were encountered on the fish silage diet.

(4) An increasing percentage of fish silage, up to 20 percent, accelerates the rate of growth. However, comparing fish silage, fish meal and a commercial protein concentrate as when fed to growing chickens, it was found that the commercial feed gave the best result but the cost of feed was also the highest. The results obtained are shown below:

Item	Weight of chicken (g)	Cost of feed (in Bahts)[1/]
20% fish silage	821.4	7
20% fish meal	912.5	8
20% commercial protein concentrate	935.2	10

9. FISH SILAGE FROM SEWAGE FISH FARMING

Tapiador (1973) made a preliminary review of sewage fish farming and its possible use as a new, non-traditional source of commercial fish meal. Production from ponds utilizing sewage water ranges from 6.5 tons/ha/year in Taiwan to 28 tons/ha/year in India (Calcutta). Tilapia harvested from sewage ponds in Taiwan weighing more than 40 g were sold for human consumption and those less than 30 g each were sold as feed for chicken, ducks and pigs. In Calcutta, the fish are grown to about 200 g in size and are sold in the market for human consumption.

The Asian Institute of Technology (AIT) is currently undertaking a research project on the utilization of sewage for fish (tilapia) farming under Professor P. Edwards. Production levels of about 20 tons/ha/year are claimed and experiments are being undertaken to dry the fish and to mix them into a fish-feed formula.

Fish produced from sewage ponds (as in India and Taiwan) and sewage streams (as in Indonesia) can be marketed for human consumption but in some areas or communities there would be an objection to eating fish produced from sewage water, not only from the aesthetic point of view but also due to possible bacterial contamination. Thus, fish produced from sewage ponds would be suitable not only for fish meal manufacture but even more so for fish silage.

The product resulting from fish silage is supposed to be sterile but experiments must be conducted on whether all harmful pathogens are destroyed or eliminated during the fish-silage process.

10. SUMMARY AND RECOMMENDATIONS

(1) The requirement of raw material for fish meal production has now apparently reached its peak and therefore the production of fish silage could provide a means of utilizing waste fish as animal feed. Waste fish is available all over the country but the quantities available are variable.

(2) The catfish and duck farmers usually buy trashfish to feed either directly or by mixing with other ingredients. These farmers could benefit by buying trashfish during glut situations and preserving them as fish silage for use when trashfish are short or when the price of trashfish is high.

(3) Some low-value fish are at present being discarded at sea; these could be converted into fish silage. A few private firms have expressed an interest in testing the commercial production of fish silage onboard their vessels.

1/ Baht 20 = U.S.$ 1

(4) The price of fish silage is comparatively cheap when compared with fish meal (about two thirds) and the fuel crisis will increase the benefits of fish silage production.

(5) The Department of Fisheries policy is to increase the freshwater and brackishwater fish production to augment fish supplies declining due to the imposition of 200-mile economic zones. The production of catfish has a target figure of 10 000 tons/year. The head, gills and offal from catfish represent about 40-50 percent. This waste could be suitable for fish silage production.

(6) The potential for producing fish silage in Thailand is good since fish waste occurs in many areas. The quantities available are small and variable. Therefore, the high investment required for fish meal production is relatively risky. Also the process of making fish silage is suitable for cottage-level operations, and requires less raw material than a fish meal plant, little capital investment is needed, little technology is involved and energy requirements are very low.

(7) More feeding trials are needed. It is hoped that commercial-scale production will start as soon as a BP silage plant which is being provided by TPI arrives in Thailand. The technical and economic feasibility of producing fish silage on a commercial scale needs to be tested. Experimental production and testing would then be introduced onboard vessels at sea.

(8) Experiments on silage produced by lactic acid fermentation should be carried out since this process has a number of advantages and Thailand has the necessary raw material.

(9) There is a promising potential for sewage fish farming and the conversion of the fish produced into fish silage should be investigated.

(10) The poultry and pig industries in Thailand can be considered as being relatively sophisticated. Therefore, if one can convince a few large feed companies to substitute fish silage for fish meal, the prospects for fish silage would be excellent.

ACKNOWLEDGEMENTS

The authors would like to express their appreciation to Mr. D.D. Tapiador, Regional Fisheries Officer and Secretary of the Indo-Pacific Fishery Commission (IPFC), FAO Regional Office for Asia and the Far East, Bangkok, for his valuable suggestions.

REFERENCES

Disney, J.G., 1979 Prospects for fish silage in Malaysia, Sri Lanka, Bangladesh and the Philippines, IPFC Occas.Pap., (1979/1):51 p.

Disney, J.G., I.N. Tatterson and J. Olley, 1977 Recent developments in fish silage. In Proceedings of the Conference on the handling, processing and marketing of tropical fish. London, Tropical Products Institute, pp. 231-40

Disney, J.G. et al., 1978 Development of a fish silage/carbohydrate animal feed for use in the tropics. Trop.Sci., 20(2):129-44

STUDIES ON THE PREPARATION OF FISH SILAGE
I. EFFECT OF QUALITY OF RAW MATERIAL AND TYPE OF ACID

K.M. Jayawardena, A. Villadsen, Q. Guneratne
and
R.G. Poulter[1]
Institute of Fish Technology, Colombo, Sri Lanka

Abstract

Fish silage was produced from good and bad quality silverbelly (Leiognathus splendens) using hydrochloric acid, formic acid and a mixture of the two. Silage produced from silverbelly held for 3 hours at 28°C initially had a rubbery texture and an even distribution of acid was difficult to achieve; this reduced the storage life of the product. Silage produced from fish held at 28°C for 12 hours had a paste-like consistency and mixing was less of a problem.

Silages prepared from silverbelly held for 3 or 12 hours at 28°C and in which the pH had been reduced to 2.5 or below by addition of hydrochloric acid remained in an acceptable condition for at least 115 days.

Silage which kept for at least 30 days, was produced from silverbelly held for 3 or 12 hours at 28°C by: (a) reducing the pH to 3.0 by the addition of hydrochloric acid; (b) by adding 0.5% formic acid and reducing the pH to 3.5 with hydrochloric acid and (c) by adding 2.5% formic acid.

1/ Seconded from Tropical Products Institute, London, U.K.

STUDIES ON THE PREPARATION OF FISH SILAGE
II. RATE OF LIQUEFACTION OF DIFFERENT TISSUES OF SILVERBELLY

K.M. Jayawardena and R.G. Poulter[1]
Institute of Fish Technology, Colombo, Sri Lanka

Abstract

Fish silage was produced from the various carcass components of good quality silverbelly (*Leiognathus splendens*) using 3.5% formic acid. Satisfactory liquefaction was found only to occur in silages containing viscera or heads. For these silages, 3.5% formic acid was sufficient to prevent spoilage for up to 2 months. However, silages made with 3.5% formic acid were found to be susceptible to mould attack, particularly on prolonged storage. Measurement of changes in non protein nitrogen content was found to be the most reliable method for assessing changes in the degree of liquefaction and autolysis in the silages. Measurement of pH change in the silage was found to give a good indication of whether the quality was deteriorating.

1/ Seconded from Tropical Products Institute, London, U.K.

STUDIES ON THE PREPARATION OF FISH SILAGE
III. DRIED SILAGE PRODUCTS

K.M. Jayawardena, Q. Guneratne, A. Villadsen and R.G. Poulter[1/]
Institute of Fish Technology, Colombo, Sri Lanka

Abstract

Fish silage was produced from low quality silverbelly (*Leiognathus splendens*): (a) by the addition of hydrochloric acid to give a pH of 2.5; (b) by the addition of 0.5% formic acid and hydrochloric acid to give a pH of 3.0; (c) by addition of 1% formic acid and hydrochloric acid to give a pH of 3.5 and (d) by addition of 2.5% formic acid. All silages kept in good condition for at least 1 month.

To make a product suitable for use in compounded chicken feed, rice bran (1:3) or maize meal (1:1) was added to the liquid silages and the mixtures dried. The resulting powders had an acceptable appearance and a pleasant odour. For the dried silage product from silage made with 2.5% formic acid and rice bran (3:1), and dried to 10.5% water content, the yield (% w/w of raw fish) was 65.3% and the protein content 32.3%.

1/ Seconded from Tropical Products Institute, London, U.K.

THE NUTRITIVE VALUE OF FISH SILAGE

As indicated in the Introduction it was difficult to separate the papers into distinct categories. For example, the papers presented in this section contain much information on the production of fish silage. However, it was felt that the papers dealing with nutritive value should be presented together as they cover the current most important aspect in the development of fish silage technology. Nine papers, including the abstract, were presented as follows:

MICROBIAL FISH SILAGE: CHEMICAL COMPOSITION, FERMENTATION CHARACTERISTICS AND NUTRITIONAL VALUE

I.P. Kompiang, Yushadi[1/] and D.C. Creswell
Centre for Animal Research and Development
Bogor, Indonesia

Abstract

Microbial silage was prepared by natural anaerobic fermentation of fresh fish mixed with various levels of molasses. The ratio of fish to molasses (w:w, fresh basis) was 100:5 (Silage A); 100:10 (Silage B); 100:15 (Silage C) and 100:20 (Silage D).

The pH of silages B,C and D decreased from approximately 7 at the time of manufacture to 4.5 on the third day and remained at or below this value during storage for 21 days. The pH of silage A was only 5.0 after 3 days of storage and this silage became spoiled by day 10. The carbohydrate content of all silages was significantly reduced during fermentation and this reduction was associated with the reduction in pH. The concentration of total nitrogen was unchanged during 21 days of storage. The soluble N (as % of total N) increased from 6 at day zero to 25 at day 21. The NH_3-N (as % of total N) also increased from 4 at day zero to 13 at day 21.

Silage C was dried and fed at varying levels of the diet to young chickens. The growth rate was compared with that obtained using traditionally prepared fish meal derived from the same batch of fish. Results indicated that the microbial fish silage was a satisfactory replacement for fish meal at levels up to 8 percent of the diet (dry fish silage basis). At very much higher levels (29 percent), growth of chickens was significantly depressed relative to the fish meal controls.

1. INTRODUCTION

Fish and fish waste can be preserved as silage, by lowering the pH either by the addition of mineral or organic acid. Recent work has also been directed toward microbiological fermentation as a means of acid preservation (Roa, 1965; Stanton and Yeoh, 1976; James *et al.*, 1977). Fermentation as a method to conserve or utilize surplus food has been widely employed in Indonesia and attempts have been made to apply this method to conserve fish or waste fish by utilizing locally available materials.

The present experiment describes the preparation of microbial silage using molasses as a carbohydrate source and the nutritional value of the product when fed to young chickens.

2. MATERIALS AND METHODS

2.1 The raw material for the fish silage, and the reference fish meal was made from the by-catch fish of shrimp trawlers (Kompiang *et al.*, this workshop).

2.2 Silage was prepared from fresh frozen fish after thawing with running water. For chemical composition studies and fermentation characteristics, the fish was put through a meat grinder with a 5 mm disc. For the feeding trial, the fish was chopped in a fish chopper. To the minced fish, molasses were added and they were mixed thoroughly and fermented anaerobically without any addition of a starter culture.

1/ Analytical Chemistry Academy, Bogor, Indonesia

3. RESULTS

3.1 Chemical composition and fermentation characteristics

The pH and visual observations were used as indicators of the course of fermentation and to determine the success of the process. The total protein content, soluble N, NH_3-N and starch content were also determined.

pH values are presented in Table 1. The pH of all silages was approximately 7 at the time of manufacture. After 3 days of fermentation the pH dropped to 4.5 and stayed at this value or even lower up to 21 days of storage, except for silage A. The pH of silage A was 5.0 after 3 days of storage and this silage had become spoiled after 10 days.

The starch content (Table 2) of all silages was significantly reduced during fermentation. The reduction in starch content was associated with, and analogous to, the reduction in pH.

The concentration of total protein (Table 3) was unchanged during 21 days storage. However the percentage of soluble N (as percent of total N, Table 4) was increased from 6 percent at day zero to 25 percent after 21 days of storage. Similarly the NH_3-N content (as percent of total N, Table 5) also increased from 37 percent at day zero to 13 percent at 21 days of storage.

Silage A was spoiled and had a foul odour at day 10 of storage. The other silages had a fresh, strong acidic smell at least up to 21 days of storage.

3.2 Nutritional value

The nutritional value of silage C was compared with fish meal prepared from the same batch of fish, by assaying the growth responses in young chickens.

A large quantity of silage C was prepared and stored for 14 days, followed by sun-drying. Composition of the fish meal control diet and the fish silage experimental diets are shown in Table 6. All diets were formulated to contain 20 percent protein, 1 percent calcium and 0.6 percent phosphorus. Each diet was fed to 40 one week-old chickens divided into 4 pens (5 males and 5 females/pen). Growth and feed intake were measured during a three-week feeding period. Body weight gain, feed consumption and feed efficiency (feed per unit gain) are presented in Table 7. The growth rate of the chickens fed diets containing silage C up to 8 percent was similar to that of the control diet. However, when a high level (29 percent) was used, a depression in growth occurred.

4. DISCUSSION

These experiments have demonstrated that fish can be preserved by natural anaerobic fermentation with molasses, i.e., without addition of any starter culture. The stability of the silage, as reported previously (Stanton and Yeoh, 1976) depends upon the fish/starch ratio. A ratio of fish to molasses 100:5 was stable for only a few days, giving off a foul odour after the tenth day. At the higher ratios, 100:10, 100:15 and 100:20, the silage was stable for at least 21 days and had a fresh acidic smell. Successful fermentation has also been reported by other workers; Roa (1965) ensiled minced herring with 10 percent molasses. Also in Australia, abalone viscera has been ensiled with 10 percent, by weight of ground malted barley; however in both cases the preparation was inoculated with L. plantarum and pH 4.5 was achieved in four days. Olley (1972) also prepared successful silages using animal waste (blood or sheep guts) and apple pulp without the addition of an inoculum. However, work in South Africa failed to produce a satisfactory silage from minced hake frames and apple pulp although the same fish material with 10 percent molasses and an inoculum was ensiled successfully. It should be noted that in the present experiment, the fish was not minced to a very fine paste, especially the material for the feedingtrial, the fish was only put through a fish chopper. The difference between the present result and the unsuccessful trial in South Africa might be due to the different area where the fish were caught, the type of fish or the type of molasses. The raw material used in Indonesia apparently contained enough bacteria for the fermentation process.

As expected, in the successful experiments, the starch content was significantly reduced during fermentation. In microbial silages, lactic acid bacteria ferment the sugar present to organic acid, thus lowering the pH. If the pH falls sufficiently low (4.5) growth of putrefactive organisms and pathogens is inhibited. This could explain the failure of silage A (fish/molasses ratio 100:5), where the pH never went below 5.0, indicating that insufficient organic acid was produced. This in turn might be due to an inadequate amount of starch. The starch content after 7 days was only 0.2 and at the same time the ammonia content was high (12.2 percent). This may also explain, at least partly, the increased pH of silage B (fish/molasses ratio 100:10) on day 21 where the starch content was already low; indeed this silage became spoiled within 2 weeks. The starch content of silage C (fish/molasses ratio 100:15) and silage D (fish/molasses ratio 100:20) remained relatively high, the pH remained stable at 4.3, and these silages could be kept even longer. Thus, it can be concluded that the success of fermentation and the storage life of the silage is dependent on the initial level of starch.

The protein content was similar throughout fermentation. The silages liquefied on day 7. The soluble N-content increased during fermentation, indicating that proteolytic enzymes are still active; after 21 days, 25 percent of the total N was in the soluble form. This contrasts with reports on chemical silage, where liquefaction occurs within a few days and the soluble N-content is 80-90 percent of total N after 14 days of storage. Conversely, the NH_3-N content of the microbial silage (11-16 percent) was much higher than in chemical silage (1.3-1.8 percent) (Kompiang <u>et al.</u>, 1979). A high level of NH_3-N has also been reported by other workers (Nillson and Rydin, 1965). The difference between microbial and chemical silage on the NH_3-N content might be, at least partly, due to the different method of storage. With microbial silage, storage is in a closed container, thus NH_3 being produced has no chance to escape, while the chemical silage is stored in open or lightly closed containers.

The feeding trial demonstrated that microbial fish silage is a good protein source which supports efficient growth of young broiler chickens at levels up to 8 percent (dry weight), similar to fish meal. This contrasts with observations on chemical silage, where at a level of 8 percent, there is a significant weight gain depression (Kompiang, unpublished data). When microbial fish silage only was used as the source of protein in the ration, as with chemical silage a significant weight gain depression is observed. However, the degree of the depression in chemical silage is larger than that with microbial silage. Thus it would seem that the nutritional value of microbial silage is better than that of chemical silage.

<u>Table 1</u>

The pH values of fish silage
(average of 6 observations)

Fish/Molasses Ratio		Fermentation period (Days)				
		0	3	7	14	21
A	100:5	6.9	5.0	5.5	-	-
B	100:10	6.8	4.6	4.5	4.5	4.8
C	100:15	6.7	4.5	4.3	4.4	4.4
D	100:20	6.6	4.4	4.3	4.3	4.3

Table 2

The starch content of fish silages (% wet weight - average of 6 observations)

Fish/Molasses Ratio		Fermentation period (Days)				
		0	3	7	14	21
A	100:5	1.7	0.4	0.1	-	-
B	100:10	3.0	1.2	0.7	0.4	0.2
C	100:15	4.0	2.2	1.3	0.7	0.6
D	100:20	5.1	3.2	2.1	1.4	1.2

Table 3

The protein content of fish silage
(% wet weight - average of 6 observations)

Fish/Molasses Ratio		Fermentation period (Days)				
		0	3	7	14	21
A	100:5	18.2	19.9	17.6	-	-
B	100:10	18.1	17.9	18.2	17.4	17.2
C	100:15	17.1	18.1	17.1	17.0	17.3
D	100:20	15.7	15.3	16.3	15.6	14.7

Table 4

Soluble N content of fish silage
(% of total N - average of 6 observations)

Fish/Molasses Ratio		Fermentation period (Days)				
		0	3	7	14	21
A	100:5	4.3	9.6	13.7	-	-
B	100:10	6.8	10.8	15.1	20.6	25.5
C	100:15	6.0	10.7	15.5	20.8	24.9
D	100:20	5.8	11.7	16.3	21.2	25.7

Table 5

NH_3-N content of fish silage
(% of total N - average of 6 observations)

Fish/Molasses Ratio		Fermentation period (Days)				
		0	3	7	14	21
A	100:5	4.9	6.5	12.2	-	-
B	100:10	4.3	7.2	8.7	11.9	16.0
C	100:15	3.8	6.2	8.3	9.7	11.6
D	100:20	2.8	6.4	6.8	10.9	14.8

Table 6

Composition of diets (%)

Ingredient	1	2	3	4	5	6
Corn	65.0	62.4	63.3	63.2	63.8	67.0
Soybean meal	25.0	32.6	29.5	27.6	25.2	-
Fish meal	8.0	-	-	-	-	-
Fish silage C	-	2.0	4.0	6.0	8.0	29.3
Dicalcium phosphate	-	0.95	0.77	0.64	0.49	-
Calcium carbonate	1.27	1.83	1.79	1.79	1.78	0.91
Common ingredients[a/]	0.72	0.72	0.72	0.72	0.72	0.72

a/ Common ingredients: vitamin/trace mineral premix 0.5%, salt .2%, santoquin .02%

Table 7

Results of feeding trials

Treatment	Weight gain (g)	Feed/consumption (g/hd/3 weeks)	Feed/gain
1. Fish meal control	646	1 149	1.76
2. 2% silage C	620	1 145	1.89
3. 4% silage C	636	1 180	1.86
4. 6% silage C	642	1 185	1.85
5. 5% silage C	645	1 193	1.85
6. 29% silage C	568	1 093	1.93

5. REFERENCES

James, M.A., K.M. Iyer and M.R. Nair, Comparative study of fish ensilage prepared by microbial fermentation and formic acid ensilage. *In* Proceedings of the Conference on the handling, processing and marketing of tropical fish. London, Tropical Products Institute, pp. 273-6
1977

Nilsson, R. and C. Rydin, Fermentation as a means of preserving organic materials. *Act. Chem. Scand.*, 17, Suppl.1:174-9
1963

Olley, J., Unconventional sources of fish protein. *CSIRO Food Preserv.Q.*, 32(2):27-32
1972

Rao, P.D., Ensilage of fish by microbial fermentation. *Fish.News Int.*, 4(3):283-6
1965

Stanton, W.R. and Yeoh Quee Lan, Low salt fermentation method for conserving trash fish under S.E. Asian conditions. *In* Proceedings of the Conference on the handling, processing and marketing of tropical fish. London, Tropical Products Institute, pp. 277
1977

NUTRITIONAL VALUE OF FISH SILAGE

I.P. Kompiang, A. Darwanto[1/] and R. Arifuddin[2/]
Centre for Animal Research and Development
Bogor, Indonesia

Abstract

Fish silages were prepared from fresh fish by its addition of 3 percent of a mixture of formic acid:propionic acid (1:1) (Silage A), 3 percent of a mixture of formic acid:propionic acid:sulphuric acid (1:0.5:2) (Silage B) or 20 percent of molasses followed by anaerobic fermentation (Silage C). Chemical composition of the silages was determined after 14 days of storage at ambient temperature. The pH values were 3.6, 2.8 and 4.3, protein concentrations (percent wet weight) were 18.0, 16.6 and 15.6, soluble N concentrations (percent total N) were 81.3, 96.4 and 21 and NH_3-N concentrations were 1.4, 2.1 and 11.0 for silages A, B and C respectively.

The nutritional properties of these silages were determined by assaying growth responses in young chickens. Whole fish meal prepared from the same batch of fresh fish was used as the control. All chicken diets contained similar levels of dry fish product (23 percent) and were made isonitrogenous. Each diet was fed to 32 one week-old chickens divided into 4 pens (5 males and 3 females/pen). Growth and feed intake was measured during a three-week feeding period. The nutritional value of silage A was similar to silage B and both were significantly ($P<0.05$) lower than fish meal or silage C. The nutritional value of silage C, while significantly better than silage A or B, was inferior to fish meal.

Chickens fed acid silages have sometimes shown symptoms of slipped tendon, which can be caused by manganase deficiency. The Mn (ppm), Ca (percent) and P (percent) levels in the dry tibia of chickens fed fish meal or silage A were 3.6, 20.9, 10.8 and 4.1, 22.6, 12.7 respectively. The level of tibial Ca and P were similar in both treatments, whereas the level of Mn was significantly ($P<0.05$) higher in the tibia of silage-fed chickens. The results indicate that the slipped tendon syndrome is not due to a lack of Mn.

1. INTRODUCTION

The aim of the present work was to compare the nutritional properties of fish silages which were prepared by the addition of different acid mixtures or by microbial fermentation. Kompiang et al. (1979) reported that the nutritional properties of fish silage prepared by the addition of 3 percent formic acid were similar to silage prepared by the addition of 3 percent of a 50:50 mixture of formic acid and propionic acid and both were inferior when compared with fish meal. Recently, Rattagool (personal communication) found that the nutritional value of fish silage prepared by addition of 3 percent of a 1:0.5:2 mixture of formic acid, propionic acid and sulphuric acid was similar to fish meal.

The nutritional properties of silage prepared from fresh fish by addition of 3 percent of a mixture of formic:propionic acid (1:1), 3 percent of a mixture of formic acid:propionic acid:sulphuric acid (1:0.5:2) or 20 percent of molasses followed by anaerobic fermentation were undertaken. The results are described below.

2. MATERIAL AND METHODS

2.1 Fish

The raw material for fish silage and the reference fish meal, was shrimp by-catch fish which was obtained as described previously (Kompiang et al., 1979).

1/ Analytical Chemistry Academy, Bogor, Indonesia
2/ Fisheries Technology Research Institute, Jakarta, Indonesia

2.2 Preparation of fish silage

Fish silage was prepared in the laboratory from the frozen by-catch after thawing in running water. The fish was chopped in a fish chopper and mixed immediately with either 3 percent (V/W) of a 1:1 (vol) mixture of 90 percent formic acid and 95 percent propionic acid (silage A), 3 percent (V/W) of a 1:0.5:2 (vol) mixture of 90 percent formic acid, 95 percent propionic acid and conc. sulphuric acid (silage B) or with 20 percent (V/W) of molasses followed by anaerobic fermentation (silage C).

The acid silages (A and B) were kept in fibreglass tanks at ambient temperature. All silages were stored for 14 days before the feed diets were prepared. Silages A and B were mixed with corn (1:1) and sun-dried while silage C was sun-dried directly.

2.3 Chemical analysis

Protein was estimated by multiplying total nitrogen, determined by the Kjeldahl procedure, by the factor 6.25.

2.4 Feeding trials

All diets contained similar levels of dry fish product and were made isonitrogenous (Table 1). Each diet was fed to 32 one week-old chickens divided into 4 pens (5 males and 3 females/pen). Growth and feed intake were measured over a three-week period.

3. RESULTS

3.1 Chemical composition

Chemical composition of the silages was determined on day 1 and after 14 days storage at ambient temperature. Prior to the addition of the acids or molasses, the pH value of the fish was 6.8.

After 14 days of storage, the pH values of silages A, B and C were 3.6, 2.8 and 4.3; protein concentrations (% wet weight) were 18.0, 17.6 and 15.6; soluble N concentrations were 81.3, 96.4 and 21.0; and NH_3-N concentrations were 1.4, 2.1 and 11. (Table 2)

3.2 Nutritional value

The nutritional value of silage A was similar to silage B and both were significantly ($P< 0.05$) lower than fish meal or silage C. The nutritional value of silage C, while significantly better than silage A or B, was inferior to fish meal (Table 3). Body weight gain (g) and feed efficiency (g feed/g gain) for silage A and B were 364, 1.96 and 358, 2.14 respectively, while for silage C they were 462, 1.90. For fish meal the results were 554 and 1.75.

Chickens fed silages A or B were noticed to have symptoms of perosis/slipped tendon, which may be caused by manganese deficiency. The Mn (ppm), Ca (percent) levels in the dry tibia of a chicken fed fish meal were 3.6, 20.9 and 10.8. For silage A the levels were 4.1, 22.6 and 12.7. The level of tibial Ca and P were thus similar in both treatments, whereas the level of Mn was significantly ($P< 0.05$) higher in the tibia of silage-fed chickens.

4. DISCUSSION

The nutritional value of silage A (with formic:propionic acid) was similar to silage B (formic:propionic:sulphuric acid) but both were inferior to fish meal. This finding contrasts with the observations of Rattagool (personal communication) who found a similar growth rate for chicken-fed fish silage or fish meal diets. This difference could be due to different fish being used, or it might be due to the period of storing the silage. Rattagool stored the silage for not more than 5 days before preparing the diets, while we stored the liquid silage for 14 days. It seems therefore necessary to explore whether reactions occuring during the storage of acidified fish generate factors which cause palatability problems or growth inhibition. Chemical oxidation of the lipid in fish silage may be different to that in fish meal. Previous work has shown that some of the toxic factors of silage are associated with the lipid fraction (Kompiang et al., 1979). This could partly explain why the nutritional value of microbial silage is better than acid silage since the former is kept anaerobically during storage, thus preventing oxidation. Another possible explanation why microbial silage is nutritionally better than acid silage is the antimicrobial substances (bacteriocide) which have been reported in silage fermentation (Wirahadikusuma, 1971).

Relatively poor weight gain has also been reported for pigs fed on fish silage absorbed into barley (Sumner, 1978) but fish silage has also been shown to be a good pig feed (Disney and Hoffman, 1976).

The symptoms of perosis/slipped tendon have been reported by other workers (Disney et al., 1977) who reported that treatment with thiamine overcame the problem. However, we were not able to confirm this finding since supplementation with thiamine did not alleviate the symptoms. The symptom of perosis/slipped tendon may also be caused by manganese deficiency, however in the present study the manganese level was similar in the diets and the manganese level in the tibia of silage-fed birds was higher than the fish meal control diet. The slipped tendon symptom observed was therefore not due to a lack of manganese.

Table 1

Percentage composition of diets

Ingredient	1	2	3	4
Corn	75.53	-	-	64.31
Fish meal	23.75	-	-	-
Silage A	-	99.28	-	-
Silage B	-	-	99.28	-
Silage C	-	-	-	34.97
Common ingredients[a/]	0.72	0.72	0.72	0.72

a/ Vitamin/trace mineral premix 5 percent, salt .2 percent, santoquin .02 percent

Table 2

Chemical composition of fish silages
(After 14 days of storage)

	Raw material	Silage A	Silage B	Silage C
pH	6.8	3.6	2.8	4.3
Protein (% wet weight)		18.0	17.6	15.6
Solube N (% total N)		81.3	96.4	21.0
NH_3-N (% total N)		1.4	2.1	11.0

Table 3

Results of feeding trials

Treatment	Body weight gain (g)	Feed consumption (g/hd/3 weeks)	Feed efficiency
Fish meal	554	970	1.75
Silage A	364	713	1.96
Silage B	358	766	2.14
Silage C	462	877	1.90

5. REFERENCES

Disney, J.G. and A. Hoffman, 1976 A dried fish silage product. In Proceedings of the Torry Research Station Symposium on Fish Silage. Aberdeen, Torry Research Station

Disney, J.G., I.N. Tatterson and J. Olley, 1977 Recent developments in fish silage. In Proceedings of the Conference on the handling, processing and marketing of tropical fish. London, Tropical Products Institute, pp. 231-40

Kompiang, I.P., R. Arifudin and J. Raa, 1979 Nutritional value of ensilaged by-catch fish from Indonesian shrimp trawlers. In Proceedings of the Torry Jubilee Conference, Aberdeen, Torry Research Station

Sumner, J., 1978 Performance of pigs fed on diets containing fish silage: evaluation in the commercial situation. Proc.IPFC, 18(3):554-8

Wirahadikusumah, S., 1969 The effect of fish silage on the quality of hens' eggs and meat of broilers. Lantbr.-Hogsk.Annlr.35, pp. 823-35

STUDIES ON THE NUTRITIVE VALUE OF FISH SILAGE FOR BROILER CHICKENS

Pong Pen Rattagool, Niracha Wongchinda and Saifon Swachathamwongratana
Fishery Technology Laboratory
Department of Fisheries
Bangkok, Thailand

1. INTRODUCTION

In 1977 the total fish catch in Thailand was 2.26 million tons. Some 40 percent of the catch is made up of trash fish and 85 percent of this is converted into fish meal. The quality of the fish meal is relatively low because the quality of the raw material is very varied. Nevertheless, the price is relatively high.

Fish silage represents an alternative means of utilizing this trash fish which could increase the revenue of the fishermen and reduce the large amounts of fish currently wasted.

One of the major difficulties of converting trash fish into fish silage, or fish meal, is the variable quality of the raw material. In many cases the fish is largely spoiled before it could be used because of the poor availability and cost of ice. The purpose of these experiments was to determine the effect of the freshness of raw material upon the production of fish silage. The optimum mixture of acids under conditions in Thailand was also investigated.

2. MATERIALS AND METHODS

Trash fish was stored with ice to preserve freshness and the fish were minced on landing. After mincing, the mince divided into three groups.

(a) to simulate fresh fish, i.e., acidified immediately

(b) to simulate average quality fish, i.e., maintained at 30°C for 8-10 hours

(c) to simulate spoiled fish, i.e., maintained at 30°C for at least 24 hours

A number of experiments were carried out on the three groups of mince.

2.1 To determine optimum acid conditions

Silage was prepared from the good, average and spoiled mince using the following acid formulations:

		good	average	spoiled
(a)	Sulphuric/formic	1.0 : 1.0 (2%)	1.0 : 1.0 (2%)	2.0 : 2.0 (4%)
(b)	Sulphuric/formic	1.5 : 0.5 (2%)	1.5 : 0.5 (2%)	3.0 : 1.0 (4%)
(c)	Sulphuric/formic	0 : 2.5 (2.5%)	0 : 3.0 (3%)	0 : 4.5 (4.4%)
(d)	Formic/propionic	0.75: 0.75 (1.5%)	1.5 : 1.5 (3%)	3.25: 3.25 (6.5%)

Changes in pH were monitored over a 22-day period and protein nitrogen (soluble and non-soluble) were estimated in the raw material and at the storage period.

In a second experiment silage made from the three types of fish mince was stored for 14 days. The acid concentrations used were as follows:

	Total	Sulphuric		Formic		Propionic
Good	2.5%	1.0	:	1.0	:	0.5
Average	5.0%	2.0	:	2.0	:	1.0
Spoiled	10.0%	4.0	:	4.0	:	2.0

Changes in pH, protein, moisture, TVB and TMA were monitored during the 14-day storage period.

2.2 Microbial fermented silage

To the three types of fish mince 30 percent molasses were added to produce a lactic acid silage. The products were stored for one month and changes in pH, protein, moisture, TVB, TMA, bacterial count and lactic acid were measured.

2.3 Feeding trials with chicken

The three types of fish mince (good, average and spoiled) were acidified with a mixture of sulphuric, formic and propionic acid (1:1:0.5, i.e., 2.5 percent). The mixtures were kept for 3 days at room temperature before being sun-dried into the following diets:

D1	Control - fish meal
D2	Fresh silage
D3	Spoiled silage
D4	Silage stored for 20 days
D5	Spoiled silage with additional vitamins
D6	Silage stored for 20 days with added vitamins
D7	Molasses silage

The final protein content of all diets was approximately 21.8 percent. Each treatment was tested with 20 chickens in duplicate kept in cages 1 x 1.5 m^2. The chickens were not sexed and they were vaccinated on day one of the trial. Body weight and food intake were checked weekly for three weeks.

3. RESULTS

3.1 The results of the first experiment to test the optimum acid concentrations are shown in Table 1 and Figure 1. It may be observed that the amount of acid required to preserve the silage varies with the quality of the raw material. Only one product containing 1.5 percent of an equal mixture of formic and propionic acids became completely spoiled. Using formic acid only (condition C) maintained a low pH throughout (approximately pH 2.0). It is significant that a mixture of sulphuric and formic acid can be used to produce a satisfactory silage even when using spoiled fish. This combination of acids is far cheaper than using formic acid alone. It is also noteworthy that a 2 percent mixture of sulphuric acid and formic acid resulted in a pH of approximately 4 with good silage, approximately 4.5 with average silage and 4.7 with spoiled silage.

The results of the second experiment, when chemical changes in the silages were measured, are shown in Table 2. The most notable results are the marked increases in TVB and TMA in the average and spoiled silages, presumably related to the bacterial activity related to the degree of spoilage.

3.2 The results of the fermented silage products are shown in Table 3. The rate of fall in pH is not known but it was presumably sufficiently rapid to preserve the product. It is significant that the pH in the spoiled silage fell slowly and did not fall below pH 5.5. The TVB and TMA levels remained fairly static after addition of the molasses.

The changes in bacterial load and lactic acid concentration are of considerable interest. Bacterial counts were high in all cases, they increased with the good and average silage but fell in the spoiled silage. Conversely the lactic acid reached the highest level in the spoiled silage sample. This is presumably due to the relative levels of lactic acid-producing and other bacteria. This aspect is worthy of further investigation.

3.3 The actual diets used in the feeding trials are shown in Table 4 and the results of the feeding trials are shown in Table 5. The gain in body weight and the feed conversion efficiencies are relatively constant, all of the diets performing as well as the control diet. In fact the product made from fresh silage produced the best growth response. It should be noted that one bird died in two of the tests.

4. CONCLUSIONS

From these experiments it can be concluded that acid and microbial silage can be produced successfully from trash fish in Thailand. In the case of acid silage, mixtures of sulphuric and formic acid can be used to reduce costs. In the case of fermented silage, no starter culture is required for successful fermentation. Successful silage products have been produced from fresh, average and spoiled fish.

Acid and microbial silages produced from fresh, average and spoiled fish have been fed to chickens and satisfactory growth responses were obtained. It is therefore concluded that fish silage can be used for feeding to poultry in Thailand.

5. ACKNOWLEDGEMENTS

The authors would like to thank Mrs. Bung-orn Saisithi, Director, Fishery Technological Development Division for her helpful suggestions. The help of Mr. Vanij Vareekul, Director of National Inland Fishery Institute and his colleague, who gave equipment and facilities to prepare the diets, was very much appreciated. The authors also wish to thank the Animal Science Agriculture Faculty, Kasetsart University where the experiments were conducted.

Table 1

The chemical composition of good, average and spoiled silages

Material	pH		Dried matter	Protein Nitrogen		
	1-9 days	10-20 days		Total N N x 6.25%	Non-soluble N N x 6.25%	Soluble N N x 6.25%
Fresh trash fish	6.9	-	24.9	18.4	15.6	2.8
Average quality of fish	7.02	-	22.7	17.5	11.9	5.7
Spoiled fish	7.65	-	16.0	17.1	7.3	10.7
Good silage						
A	3.9	4.0	26.4	17.8	5.1	12.8
B	3.6	3.7	26.4	17.6	4.4	13.2
C	2.0	2.2	26.1	17.1	4.6	12.6
D	4.9	5.0	25.9	18.4	5.5	12.8

Table 1 (Cont.)

The chemical composition of good, average and spoiled silages

Material	pH 1-9 days	pH 10-20 days	Dried matter	Protein Nitrogen: Total N N x 6.25%	Protein Nitrogen: Non-soluble N N x 6.25%	Protein Nitrogen: Soluble N N x 6.25%
Average silage						
A	4.4	4.6	24.7	17.7	5.8	11.8
B	4.3	4.4	25.5	17.8	6.7	11.0
C	2.2	2.3	27.0	17.4	6.5	10.9
D	4.8	4.8	23.5	18.3	6.6	11.6
Spoiled silage						
A	4.7	4.8	22.0	18.3	7.0	11.3
B	4.6	4.6	23.1	18.0	6.7	11.2
C	2.0	2.1	23.8	15.6	4.9	10.7
D	4.7	4.7	18.8	17.7	5.7	12.0

Table 2

Chemical properties of good, average and spoiled silage (Acid Silage)

Material	Storage (day)	pH	% Total protein based on (wet)	% Total protein based on (dry)	Moisture (%)	Ash	Calcium	Phosphorous (%)	TVB mgN/100 g sample	TMA mgN/100 g sample
Good fish	0	6.8	17.16	74.64	77.01	3.75	-	0.72	90.0	30.0
Food silage	0	2.5	17.46	70.03	75.07	4.61	-	-	50.0	8.0
	7	3.8	17.46	69.2	74.77	-	0.18	0.40	42.8	14.0
	14	3.8	19.31	65.04	74.31	-	-	-	90.0	18.0
Average fish	0	7.2	17.50	76.09	76.90	4.34	-	0.78	138.0	40.0
Average silage	0	1.4	17.17	69.92	74.41	4.65	-	-	102.0	32.8
	7	2.5	16.77	68.01	73.96	-	0.23	0.36	110.0	34.0
	14	2.6	17.71	67.39	76.72	-	-	-	120.0	40.0
Spoiled fish	0	7.65	18.50	88.05	78.99	4.77	-	0.81	382.0	44.0
Spoiled silage	0	1.45	16.32	62.84	74.05	6.01	-	-	260.0	30.0
	7	2.1	16.23	69.54	73.72	-	0.26	0.29	266.0	28.8
	14	2.15	17.06	65.39	73.91	6.7	-	-	344.0	48.0

Fig. 1

pH changes in good, average and spoiled silages.

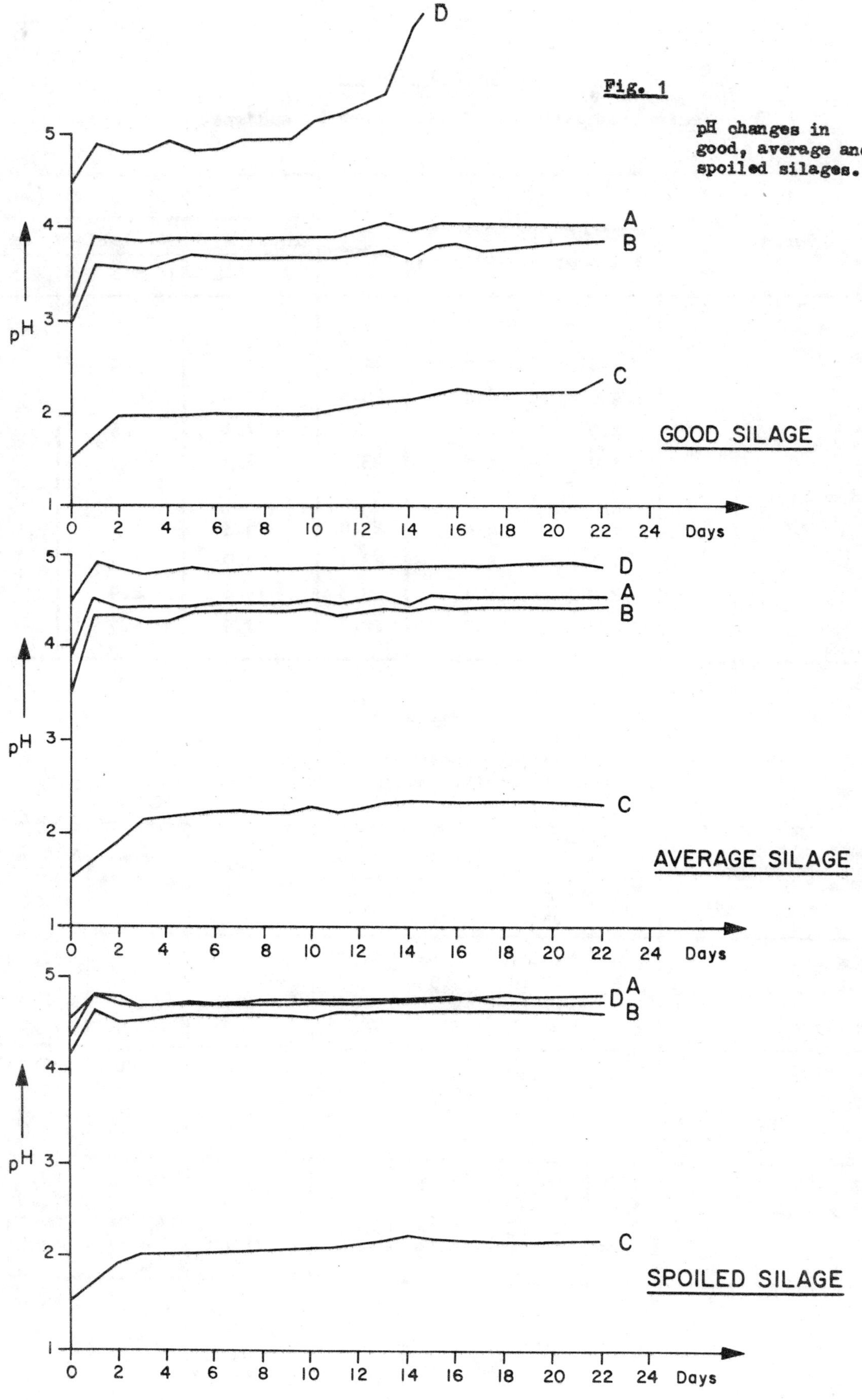

Table 3

Changes in good, average and spoiled silages with 30% molasses

Material	Storage (days)	pH	% Protein Liquid sample	% Protein Dried sample	Moisture (%)	TVB gN/ 100 g sample	TMA gN/ 100 g sample	TPC # colonies per g of sample	Lactic acid (%)
Good fish	0	6.80	17.16	74.64	77.01	90.0	30.0	4.8×10^{8}	-
Good silage	0	5.85	14.68	46.87	68.68	-	-	5.3×10^{7}	-
	7	4.25	14.77	53.59	72.49	150.8	18.4	2.4×10^{9}	0.64
	14	4.30	16.91	60.00	71.82	220.0	26.4	7.1×10^{10}	0.66
	28	4.35	-	-	-	240.0	22.0	1.5×10^{10}	0.72
	32	-	-	-	-	280.0	24.0	-	0.65
	42	-	-	-	-	-	-	-	0.79
Average fish	0	7.20	17.50	75.75	76.90	138.0	40.0	1.0×10^{9}	-
Average silage	0	5.90	15.45	45.22	65.83	-	-	1.0×10^{9}	-
	7	4.27	15.66	49.20	68.17	214.0	32.0	1.1×10^{9}	0.63
	14	4.25	16.40	52.30	68.64	334.0	38.0	4.2×10^{10}	0.64
	28	4.25	-	-	-	204.0	34.0	100×10^{10}	0.70
	32	-	-	-	-	316.0	36.0	-	0.72
	42	-	-	-	-	-	-	-	0.94
Spoiled fish	0	7.65	18.50	88.05	78.99	382.0	44.0	5.0×10^{8}	-
Spoiled silage	0	6.70	16.10	49.88	67.72	-	-	6.3×10^{8}	-
	7	5.75	15.51	53.67	71.10	312.0	40.0	6.0×10^{7}	0.65
	14	5.7	17.56	60.89	71.16	378.0	58.8	1.5×10^{6}	0.70
	28	5.5	-	-	-	240.0	38.0	8.0×10^{6}	1.26
	32	-	-	-	-	362.0	42.0	-	0.69
	42	-	-	-	-	-	-	-	1.58

Table 4

The composition of diets for the feeding trials

Ingredient	(Control) D_1	Composition of diet (% w/w)					
		D_2	D_3	D_4	D_5	D_6	D_7
Fish meal	10.0	-	-	-	-	-	-
Fish silage	-	10.0	10.0	10.0	10.0	10.0	10.0
Soy meal	20.0	20.0	20.0	20.0	25.0	20.0	20.0
Corn meal	65.0	65.0	65.0	65.0	60.0	64.0	64.0
Bone meal	1.0	1.0	1.0	1.0	1.0	1.0	1.0
Salt	0.5	0.5	0.5	0.5	0.5	0.5	0.5
Alif Alif	2.5	2.5	2.5	2.5	2.5	2.5	2.5
Vitamin-mineral	1.0	1.0	1.0	1.0	1.0	2.0	2.0
Total (kg)	100.0	100.0	100.0	100.0	100.0	100.0	100.0
% Final protein	21.18	21.86	21.83	21.08	21.15	20.77	20.57

Table 5

Results of feeding trials

Week	(Control)	Diets					
	D_1	D_2	D_3	D_4	D_5	D_6	D_7
(1) Body weight (g)	88.13	91.10	87.16	96.41	85.63	86.25	83.13
Feed conversion	1.74	1.77	1.74	1.57	1.56	1.72	1.89
(2) Body weight (g)	190.63	189.07	190.94	219.00	185.63	183.70	199.40
Feed conversion	1.87	1.89	1.83	1.64	2.08	1.82	1.77
(3) Body weight (g)	298.75	338.73	317.50	328.00	302.50	312.50	301.30
Feed conversion	2.02	1.89	2.00	2.04	2.14	1.93	2.13
Total feed intake (g)	562.50	564.80	552.70	573.70	563.40	525.60	552.80
% chick's mortality	0	3.1	0	3.1	0	0	0

(Note 3.1% equals one bird)

FISH SILAGE IN THAILAND: FEEDING TRIALS ON BROILER CHICKEN

Pong Pen Rattagool, Niracha Wongchinda and Saifon Swachathamwongratana

Fishery Technology Laboratory
Department of Fisheries
Bangkok, Thailand

1. INTRODUCTION

In 1977 some 836 600 tons of trash fish were landed in Thailand. The majority is converted into fish meal by the 74 existing fish meal plants, but large quantities are wasted. Many of the fish meal plants are old and the quality of the meal is poor but the price of the meal is still high. Fish silage presents a possible alternative to fish meal as the equipment required is simple, the quantities of raw material required are flexible and the process could well be more economic than fish meal production.

Earlier work in Thailand (Rattagool et al., in press) has shown that mixture of sulphuric acid and formic acid can preserve fish for one week without problems of putrefaction and mould growth provided the mix is stirred daily. Also in Thailand, liquid silage when mixed with corn meal or rice bran (1:1) can easily be sun-dried in 1 or 2 days. The resulting product will keep for long periods provided the moisture content is below 10 percent.

2. MATERIALS AND METHODS

Good quality scad (*Decapterus russelli*) was purchased in the vicinity of Bangkok. The fish was divided into two groups; the first group was cooked for 15-20 minutes before adding acid and the second group was minced and 2 percent of an equal mixture of sulphuric and formic acids were added. The silage was then stored in plastic containers at a room temperature of approximately 30°C.

A series of chicken diets were made up using fish meal made locally from the same raw material, the two fish silage products and a combination of other ingredients such as corn meal, vitamin premix, etc. The composition of these diets was as follows:

- D1. 5 percent fish meal
- 2. 10 percent fish meal
- 3. 5 percent dried fish silage
- 4. 10 percent dried fish silage
- 5. 5 percent dried, cooked silage
- 6. 10 percent dried, cooked silage
- 7. Commercial poultry feed
- 8. Basal feed excluding any fish
- 9. 15 percent wet silage
- 10. 30 percent wet silage

The actual combination of ingredients in these diets is shown in Table 1. The chemical composition of the diets was also analysed for protein, fat, ash and moisture content. The methods used were as follows:

Total nitrogen content was determined by the macro-Kjeldahl method. The pH was measured directly in the semi-solid material with a Corning pH meter. Ash content was determined by ignition of the dry matter and heating in a muffle furnace at 600°C. Fat content was measured by Soxhlet extraction with petroleum spirit (60-80°C) for at least 8 hours.

For the poultry feeding trials, one day unsexed broiler chickens were obtained from a commercial producer. For each of the ten dietary treatments these replications of 12 birds each were used. The birds were weighed, vaccinated and each group was kept in cages maintained under constant conditions.

The food intake and body weight gain were measured weekly and 1 percent of vitamin premix was applied in two stages, i.e., up to 14 days starter levels were used. Thereafter, finisher levels were used (up to 56 days).

3. RESULTS

The chemical composition of the ten dietary treatments are shown in Table 2. It may be observed that an attempt was made to make the diets iso-nitrogenous and the analysis showed that the diets were relatively constant in protein and energy levels.

The results of the feeding trials are shown in Table 3. The relatively high level of mortality is difficult to explain and gives cause for concern, but as deaths were high with the commercial poultry diet it does not appear that the deaths were due to palatability problems or toxic factors in the silage products.

No statistical analysis of the data was carried out but it appeard that the dried silage products gave a growth response similar to fish meal but the best response was obtained with the commercial diet. There was little to choose between the feed conversion efficiency of the fish meal and fish silage diets; there also seemed to be little difference between the dried silage product and the dried, cooked silage product.

Assuming that there was little difference in performance between the fish meal, fish silage and boiled fish silage diets, the cost of producing these feeds was calculated. The cost of the ingredients are shown in Table 4 and the comparative cost of the various diets are shown in Table 3. Comparing the diets containing 5 percent fish meal or 5 percent fish silage, it was calculated that the fish meal diet was 13.7 percent higher, while the weight gain was little affected. It was therefore concluded that the use of fish silage provides a more economic means of utilizing trash fish than converting it into fish meal.

4. CONCLUSIONS

These experiments indicate that the production of fish silage using a combination of sulphuric and formic acids provides a viable alternative to the production of fish meal. This is of particular interest in Thailand as some fish meal plants are closing due to inadequate supplies of fish and the high cost of fuel. The introduction of fish silage production could also be of benefit to the large number of low-income fishing families.

The results obtained also suggest that even if chicken fed with silage diets have a reduced weight gain when compared to commercial or fish meal diets, the lower cost of production argues in favour of introducing the use of fish silage.

5. ACKNOWLEDGEMENTS

The authors would like to thank Miss Supaporn Isariyodom for her valuable help (Department of Animal Science, Kasetsart University) and their colleagues for help with the chemical analysis.

6. REFERENCE

Ratagool, P., N. Wongchinda and S. Surachathamrongratana, Fish silage production in Thailand. (in press)

Table 1

The combination of ingredients in the diets for broiler chicken

Ingredient	Diets									
	1	2	3	4	5	6	7	8	9	10
Fish silage	-	-	5.0	10.0	5.0	10.0	COMMERCIAL	-	15.0	30.0
Fish meal	5.0	10.0	-	-	-	-		-	-	-
Soy meal	28.0	22.0	27.0	20.5	27.0	20.5		34.0	30.0	26.0
Corn meal	62.0	63.0	63.0	64.5	63.0	64.5		61.0	50.0	39.0
Alif Alif	2.5	2.5	2.5	2.5	2.5	2.5		2.5	2.5	2.5
Bone meal	1.0	1.0	1.0	1.0	1.0	1.0		1.0	1.0	1.0
Salt	0.5	0.5	0.5	0.5	0.5	0.5		0.5	0.5	0.5
Vitamin & mineral	1.0	1.0	1.0	1.0	1.0	1.0		1.0	1.0	1.0
Total (kg)	100.0	100.0	100.0	100.0	100.0	100.0		100.0	100.0	100.0
Final Prot. (% w/w)	20.3	19.3	21.1	19.6	20.8	20.9	19.1	22.5	22.3	22.2
Food/Baht/kg[a]	4.02	3.78	3.47	3.29	3.49	3.31	5.00	3.67	3.38	3.08

Table 2

Chemical composition of the diet

Treatment (or Diet)	Protein (%)	Fat (%)	Ash (%)	Moisture (%)	Energy cal/kg
1	22.3	3.9	6.5	10.2	122.08
2	19.3	2.6	6.6	11.1	106.26
3	21.1	3.4	5.3	11.7	120.81
4	19.6	3.8	5.6	9.7	118.55
5	20.8	2.4	4.9	12.2	110.64
6	20.9	3.9	5.8	10.5	125.04
7	19.1	2.3	6.5	10.8	102.52
8	22.5	2.0	5.1	10.8	114.78
9	22.3	5.5	6.5	20.5	145.06
10	22.2	4.7	6.6	30.3	137.03

Energy in Cal/kg food = percent fat x 9.02 + percent crude protein x 4.27

Table 3

Results of 8-week feeding trial (36 chickens per diet)

	Diets									
	1	2	3	4	5	6	7	8	9	10
Percent chicken death	13.9	8.3	13.9	11.1	30.5	2.8	13.9	11.1	16.7	16.7
Initial weight (g)	40.9	40.1	41.4	39.9	40.9	40.9	40.1	39.0	40.0	39.7
Final weight (g)	1400	1375	1356	1398	1354	1434	1604	1064	1433	1533
Food intake (kg)	3.14	3.20	3.15	3.29	3.44	3.24	3.40	2.88	3.87	4.56
Feed conversion ratio	2.24	2.33	2.32	2.36	2.54	2.26	2.12	2.70	2.70	2.98
Food price (Baht/kg)[a/]	4.02	3.98	3.47	3.29	3.49	3.31	5.00	3.67	3.38	3.08

Table 4

The price of ingredients

Ingredients	Price (Baht/kg)[a/] March 1978
Trash fish	1.25
Maize meal	2.50
Soy meal	5.70
Fish meal	7.40
Salt	0.50
Bone meal	3.00
Acacia (Alif Alif)	1.90
Vitamin	25.00

a/ 20 Baht = U.S.$ 1

FISH SILAGE IN THAILAND: FURTHER TRIALS ON BROILER CHICKEN

Pong Pen Rattagool, Saifon Swachathamwongratana and Niracha Wongchinda
Fishery Technology Laboratory
Department of Fisheries
Bangkok, Thailand

1. INTRODUCTION

The work reported in this paper is an extension of the work reported in the previous paper, but additional diets were investigated. The effect of the lipid content of fish silage upon the growth of chicken was also investigated. The fish silage was again produced using a combination of sulphuric and formic acids. The sulphuric acid is not only three times cheaper than formic acid but may also cause less damage to vitamin B_{12} in the diet. Some propionic acid was also used to control mould growth.

The trash fish in Thailand, which forms the potential raw material for silage production, consists of a mixture of species, depending upon the fishing area, the season and the size of the vessel. The trash fish was divided into 6 groups and then chemical composition was determined.

2. MATERIALS AND METHODS

The trash fish was caught by trawler (15-50 ton capacity) and the quality was relatively good. Both fish silage and fish meal were produced from the same batch of fish. The fish meal was made under normal commercial conditions and the stickwater was not separated. The fish silage was produced from minced or chopped fish which were acidified with 2.5 percent of a mixture of sulphuric, formic and propionic acids (1:1:0.5). The product was well mixed and stored for 3 days at ambient temperature ($30^{\circ}C \pm 3^{\circ}C$) prior to mixing with the various diet ingredients which were then sun-dried in one or two days.

The diets were prepared to contain fish meal, fish silage, boiled fish silage and fish silage from which the lipid had been extracted with acetone. There were eleven diets as follows:

Diet		
1	5%	fish meal
2	5%	fish meal with lipid extracted
3	5%	fish meal plus lipid extracted for silage
4	5%	fish meal with lipid extracted but with silage lipid extract added
5	20%	fish meal
6	5%	dried fish silage
7	5%	silage with lipid extracted
8	20%	dried fish silage
9	20%	boiled fish silage
10		Commercial poultry feed
11		Basal diet including vitamins and minerals but no fish meal or silage

The diets were all made iso-nitrogenous as far as possible to give a protein content of approximately 20 percent. The composition of the fish, silage, fish meal and other ingredients were analysed for ash, fat, moisture, total nitrogen, pH, calcium and phosphorus contents. Ash content was determined by ignition of the sample followed by heating overnight in a muffle furnace at $600^{\circ}C$. Fat content was measured, after drying at $105^{\circ}C$, by Soxhlet extraction using petroleum spirit ($60-80^{\circ}C$) for at least 8 hours. Moisture was measured by oven-drying overnight at $105^{\circ}C$. Protein content was measured by the macro-Kjeldahl method and pH was measured with a Corning pH meter. Phosphorous was determined by the phosphomolybdate method and calcium determination was performed by nitric acid digestion followed by atomic absorption.

The feeding trials were conducted as reported previously (Rattagool, 1979), i.e., three replicates of 12 birds per diet treatment, and they lasted for 8 weeks.

3. RESULTS

The chemical composition of the groups of trash fish and the various ingredients of the diets are shown in Tables 1 and 2 respectively. The shrimp by-catch consists predominantly of two species, namely silver belly and threadfin bream. It is also interesting to note that the overall lipid content of the by-catch is relatively low (1.1 according to Table 2). The lipid content of the dried silage is also not significantly higher than that of fish meal.

The various components in the poultry chicks are shown in Table 3 and the chemical composition of the diets are shown in Table 4. The protein content of the various diets are relatively constant (around 20 percent) with the exception of basal diet, presumably due to the high level of soy meal. The total lipid content in the diets does reflect the levels of fish lipid, e.g., diets 2 and 7 have low lipid contents. However, there is little variation in the total lipid content and in fact the commercial and basal diets have the highest lipid content. Nevertheless, it is the level of fish lipid in the diet which is under investigation.

The results of the feeding trials are shown in Table 5. Although the best weight gain and feed conversion efficiency were obtained with the commercial diet the growth response from the remaining diets was not markedly different. However, the weight gain with 5 percent silage was less than that resulting from 5 percent fish meal. Taking the 5 percent fish meal diet as the control, the relative growth resulting from the remaining diets was as follows:

Control 1	100.0%
2	95.1%
3	93.9%
4	99.8%
5	100.0%
6	90.1%
7	87.0%
8	97.4%
9	100.1%
10	106.2%
11	81.9%

This analysis of the results indicates that the growth obtained with fish meal is marginally superior to that resulting when fish silage is used (i.e., of diets 1 and 6). However, the results can only be said to be inconclusive as far as the effect of fish lipid content is concerned. No clear pattern emerges and certainly the removal of lipid from silage did not result in an increased growth response. It should be pointed out that the silage used in these experiments is stored for only 3-5 days prior to sun-drying. Thus, the problems of lipid oxidation reported by Kompiang (personal communication) when silage is stored for one month would not be expected under the present experimental conditions.

To investigate the likelihood of fish taint in the poultry reared on fish silage, taste panel tests were carried out. Breast and thigh meat was sealed in plastic bags and steam heated for 10 minutes. The samples were then scored for texture and flavour by 7 panel members. The meat was found to be tender and juicy and no silage flavour was reported although one panel member did comment on a mild silage odour on one of the larger, fatter, birds.

4. ACKNOWLEDGEMENTS

The authors would like to thank Mrs. Bung-orn Saisithi, Director of the Fish Technology Development Division for her valuable suggestions and encouragement. They would also like to thank Miss Attaya Kungsuwan for assistance with the chemical analysis and the Animal Science Faculty of Kasetsart University who provided the facilities for the feeding trials.

5. REFERENCE

Ratagool, P., N. Wongchinda and S. Surachathamrongratana, Fish silage production in Thailand. (in press)

Table 1

Proximate composition of fish (trash fish) (initial pH = 6.2)

Material	% of trash fish	Protein (%)	Lipid (%)	Moisture (%)	Ash (%)	Calcium (mg/100 g)
(1) Silver belly (*Leiognathus* spp.)	25.0	15.4	1.5	73.0	7.2	2630
(2) Sixtooth threadfin (*Nemipterus* spp.)	24.0	19.9	3.4	72.3	4.6	1137
(3) Sardine (*Clupea* spp.)	10.0	18.4	5.0	76.4	1.0	-
(4) Lizard fish (*Saurida* spp.)	9.0	19.9	0.3	75.6	5.4	1738
(5) Scad (*Caranx* spp.)	5.0	21.4	1.4	73.4	3.1	704
(6) Miscellaneous	27.0	15.9	-	79.1	3.5	-
Trash fish	100.0	17.6	2.3	74.8	5.6	900.57

Table 2

Proximate composition of diet ingredients

	Protein (%)	Moisture (%)	Fat (%)	Ash (%)	Phosphorous (%)	Calcium (mg/100 g)	Crude fibre	Energy (cal/kg)
Trash fish	18.0	77.1	1.1	6.3	0.79	1020	-	86.47
Fish silage (wet)	14.1	75.7	1.0	7.3	-	-	-	69.27
Boiled fish silage (wet)	15.1	75.3	1.5	7.3	-	-	-	78.14
Dried fish silage	52.6	11.0	8.2	28.6	0.95	11600	0.32	299.02
Fish meal	55.0	10.5	7.2	31.9	2.49	12900	0	299.48
Soy meal	46.3	11.0	7.6	7.0	0.62	443	6.26	266.69
Cornmeal	11.0	10.3	4.6	5.5	0.21	27.3	1.57	88.51
Alif Alif	15.0	6.0	2.0	31.5	0.12	1800	22.88	81.73

Table 3

Composition of diets

Ingredient	(kg) Ingredients in diet (% w/w as fed basis)										
	1	2	3	4	5	6	7	8	9	10	11
Fish meal	5.0	5.0	5.0	5.0	20.0	-	-	-	-	PRIVATE CO.	-
Fish silage	-	-	-	-	-	5.0	5.0	20.0	20.0		-
Soy meal	22.0	22.0	22.0	22.0	2.0	22.0	22.0	2.0	2.0		29.0
Corn meal	68.0	68.0	68.0	68.0	73.0	68.0	68.0	73.0	73.0		66.0
Alif Alif	2.5	2.5	2.5	2.5	2.5	2.5	2.5	2.5	2.5		2.5
Bone meal	1.0	1.0	1.0	1.0	1.0	1.0	1.0	1.0	1.0		1.0
Vitamin-mineral	1.0	1.0	1.0	1.0	1.0	1.0	1.0	1.0	1.0		1.0
Salt	0.5	0.5	0.5	0.5	0.5	0.5	0.5	0.5	0.5		0.5
Crude protein (%)	19.3	18.8	19.2	18.7	20.5	19.6	18.7	18.8	18.5	19.2	29.4

Table 4

Proximate composition of diets

Diet	Protein (%)	Moisture (%)	Fat (%)	Ash (%)	Phosphorous (%)	Calcium (%)	Crude fibre (%)	Energy (cal/kg)
1	19.3	11.1	4.3	6.5	0.5	970	4.9	121.2
2	18.8	10.6	3.5	6.4	0.5	3000	4.7	111.0
3	19.1	9.1	5.0	6.3	0.5	1400	4.9	127.3
4	18.7	10.3	4.5	6.4	0.5	1800	3.9	120.9
5	20.5	20.4	4.2	9.4	0.8	3300	4.9	125.5
6	19.6	9.8	3.3	6.9	0.5	1100	4.4	1113.4
7	19.7	9.5	2.7	6.4	0.5	2400	4.9	108.2
8	18.8	8.9	3.5	9.2	0.6	5200	3.3	111.8
9	18.5	11.1	3.6	9.7	0.6	3100	2.5	111.9
10	20.4	9.6	7.4	5.7	0.4	641	3.9	153.5
11	19.2	10.4	8.2	7.3	0.7	1800	3.5	155.9

Table 5

Results of feeding trial

	1	2	3	4	5	6	7	8	9	10	11
4 Weeks											
Weight (g)	534.3	514.4	528.3	553.0	623.6	541.2	500.6	538.4	565.0	667.9	461.2
Feed conversion efficiency	2.34	2.26	2.21	2.18	2.04	2.22	1.14	1.00	1.84	1.75	2.29
Feed intake (g)	1159.5	1076.9	1085.6	1124.2	1193.7	1119.2	1035.4	1002.2	970.8	972.2	1105.6
8 Weeks											
Weight (g)	1507.2	1434.1	1415.1	1503.9	1584.1	1358.9	1311.4	1543.7	1520.8	1601.2	1272.8
Feed conversion efficiency	2.58	2.66	2.48	2.47	2.41	2.70	2.72	2.17	2.18	2.11	2.82
Feed intake (g)	3789.6	3711.2	3417.4	3618.9	3732.1	3574.8	3462.3	3264.7	3244.2	3579.7	3308.3

STUDIES ON FISH SILAGE IN SRI LANKA - A SUMMARY

R.G. Poulter[1] and K.M. Jayawardena
Institute of Fish Technology, Colombo, Sri Lanka

and

P. Ganegoda and K.N.P. Ranaweera
Veterinary Research Institute, Peradeniya, Sri Lanka

1. INTRODUCTION

The acid ensilage of fish to produce a protein supplement for feeding to animals is not a new process. The basic technology was developed more than 50 years ago and substantial quantitites of silage are produced in countries such as Denmark and Poland, mainly for feeding to pigs. In Sri Lanka, fish silage was first produced experimentally by Gunasekera and Lantz in 1955.

In recent years, however, there has been renewed interest in fish silage as it represents a means of utilizing waste fish and offal in situations where conventional fish meal production is inappropriate.

Several institutes in the IPFC region are actively involved in projects to develop fish silage processes and this paper briefly summarizes the work being undertaken in Sri Lanka by the Institute of Fish Technology (IFT) and the Veterinary Research Institute (VRI).

2. FISH RESOURCES

There are two potential sources of raw material for silage production in Sri Lanka. One is the by-catch of prawn trawlers operating in the Mannar/Jaffna area which consist mainly of silver belly (*Leiognathus splendens*). Although small quantities of these fish are landed and sold for human food, the majority is either discarded at sea or converted into fish meal using crude processing techniques which result in a loss of at least 25% of the protein.

The other potential raw material for silage production is prawn processing waste. There are 15 processing plants in the Colombo area and most of the 3 to 4 tons per day of prawn waste produced is simply removed to the municipal rubbish dump. There is also a limited quantity of waste fish offal produced at the main fish market in Colombo.

3. LABORATORY STUDIES ON SILAGE PRODUCTION

A number of studies have been carried out to determine the optimun conditions for producing liquid silage from silver belly and prawn waste. The use of hydrochloric and formic acids have been considered together with the effect of fish quality. A summary of the results is given below. Full details of the work are to be published shortly in a series of papers in the Bulletin of the Fisheries Research Station, Sri Lanka.

Liquid silage made from silver belly, which kept for at least one month, could be produced either by reducing the pH to 3.0 by the addition of hydrochloric acid or by adding 2.5% of formic acid. Surprisingly, it appeared that silage made from low quality fish remained free from obvious spoilage for longer than silage made from good quality fish. This seemed to be related to the ease with which the acid and minced fish could be mixed during the early stages of ensiling, i.e., with partially spoilt fish the penetration of the acid was more rapid. Silage produced with 3.5% of a 1:1 mixture of formic and propionic acids was found to keep for at least 6 months, probably due to the prevention of mould growth by the

---

1/ Tropical Products Institute, London, U.K.

propionic acid. Changes in the pH of the silage were found to be a good indication of quality deterioration and measurement of the non protein nitrogen content was found to be the best means of measuring autolysis/liquefaction. In an experiment designed to measure the rate of liquefaction in different parts of silver belly ensiled with 3.5% formic acid it was found that satisfactory liquefaction occurred only when viscera or heads were present.

Dried silage products with an acceptable appearance and pleasant odour were produced by drying the mixture obtained by adding either rice bran or maize meal to liquid silver belly silage prepared with either hydrochloric or formic acid. For the dried silage produce made from silver belly silage with 2.5% formic acid and rice bran (3:1), and dried to 10.5% water content, the yield was 65.3% and the (% w/w of raw fish) protein content 32.3°/oo.

Generally, the pH of liquid silage made from prawn waste was less stable than that made from silver belly and hence it was more susceptible to spoilage. However, silage, which kept for at least one month, could be produced by maintaining the pH at 2.5 with hydrochloric acid or by adding 2% formic acid and maintaining the pH at 3.5 with hydrochloric acid. It should be noted that for prawn waste larger quantities of hydrochloric acid are required to give the desired pH. Addition of 3.5% formic acid alone initially reduced the pH of prawn waste silage to 4.0, the silage remained in good condition for 2 weeks but then the pH began to increase rapidly. After 3 weeks of storage the pH had reached 5.0 and the silage had become spoiled. As expected, the chitinous part of the prawn waste did not break down during the ensilage process.

Silages were produced from mixtures (1:1 up to 10:1) of prawn waste and silver belly using 3.5% formic acid. With high proportions of similar belly present the silages remained free of obvious spoilage for up to 3 weeks but as the ratio of prawn waste to silver belly increased so did the tendency of the silage to spoil.

Liquid produced from prawn waste alone and prawn waste (3 parts) with fish waste (i.e., heads, viscera etc. - 1 part) were mixed with rice bran (1:3, rice bran to silage) and dried. The appearance and odour of both the dried products were acceptable. The final water contents were 14% and the true protein content (i.e., total N minus chitin N x 6.25) was found to be 21% for the prawn waste sample and 23% for the prawn/fish waste sample.

As a result of the experience gained in these laboratory studies a number of recommendations were formulated for pilot-scale comercial trials in Sri Lanka.

(1) Formic acid rather than hydrochloric acid should be used for silage production since it gives more consistent results and is readily available.

(2) On an experimental scale a stable silver belly silage can be produced using 2.5% formic acid but for commercial production it is advisable to add sufficient formic acid to reduce the pH to below 4. For silver belly held for up to 24 hours without ice this requires the addition of 3.5% formic acid. For prawn processing waste concentrations of formic acid of 4% or above will be required.

(3) As liquid silage made with formic acid may be susceptible to mould attack on prolonged storage, it should utilized or dried as soon as possible after production. If the product has to be stored in the wet state it should be stirred regularly and any rise in pH remedied by addition of further quantities of formic acid. If persistant mould attack proves to be a problem the addition of low concentrations of propionic acid should be considered.

(4) On economic grounds rice bran rather than maize meal should be used as a filler material in the production of dried fish silages. Only good quality rice bran (i.e., protein content 10% or above) should be used. The optimum ratio of rice bran to silage is 1:3.

4. POULTRY FEEDING TRIALS

Imported fish meal and meat meal are the principal source of animal protein used to supplement coconut meal and cereal-based rations for non-ruminant farm animals and poultry in Sri Lanka. However, in view of the poor quality and price of these feedstuffs, it is imperative that optimum use be made of the locally available fish caught as a by-catch from prawn fishing and fish offal. Research to date has shown that it is cheaper and less involved to convert these raw materials into fish silage rather than into fish meal. However, it is not clear whether fish silage can successfully replace fish meal or meat meal in the safe and economical production of human food from livestock. A series of feeding experiments and field tests was therefore planned to evaluate dried fish silage as an ingredient in the practical diets of broiler and layer chicken. Some of these experiments and field tests are still underway; the results of those that have been completed are summarized below.

A dried fish silage product, prepared by adding maize meal to a liquid oil sardine silage produced with formic acid, had three parts of maize to one of fish on a dry weight basis and contained 26% crude protein. When this material replaced (on a weight for weight basis) part or all of the average quality fish meal (52% crude protein) incorporated in broiler starter and finisher diets at 20% level, the weight gain of the test birds was found to be marginally superior, provided the crude protein deficits in the test diets were made good by the inclusion of soybean meal at the expense of some rice bran and maize. Their feed conversion efficiencies were also significantly better, presumably due to an overall improvement in the metabolizable energy contents of the test diets.

A dried fish silage produced from average quality silver belly using formic acid and rice bran as filler material, had 0.85 parts of fish to one of rice bran on a dry weight basis and contained 35.5% crude protein. The performance of broiler chicken, as judged by the weight gain and feed conversion efficiency, was not adversely affected when this material replaced part or all of the meat meal (crude protein content 50%) incorporated into control diet at a level of up to 21%, provided that the dietary protein deficit caused by the replacement was corrected as in the previous experiments. Indeed, in the starter stage the feed conversion efficiencies for the silage treatments were significantly superior to the control.

The fish by-catch from prawn trawlers (i.e., silver belly) is rarely stored on ice and hence it is usually of poor quality by the time it reaches the processing point. An experiment was therefore carried out to determine the effect of fish spoilage on the feeding value of dried fish silage products. Batches of silver belly stored for 0 hours (very fresh), 12 h (average) and 24 h (spoiled) at 30°C were separately ensiled with formic acid, mixed with rice bran, and sun-dried. Quite unexpectedly, these three products were almost indistinguishable from one another, when evalated by feeding them to broiler chickens at 20% and 16% dietary levels during the starter and finisher stages, respectively. However, a laboratory produced fish meal (61% crude protein), prepared from the same fish species under carefully controlled conditions of cooking and drying, proved to have a significantly higher nutritive value.

Perhaps the most interesting observation made in this third experiment was the occurrence of a leg weakness in those birds that were fed the dried silage products. The extent of the weakness ranged from stiff gait through limping to clear-cut perosis. It is noteworthy that in the first two experiments, birds fed diets containing dried silage products did not develop any leg weakness. The supplementary vitamin levels used in these experiments were in excess of the normal requirements for chicken whereas in the last experiment, the supplementary vitamins and trace minerals were at the normal recommended level. Moreover, the broiler chicken used in the latter trial were a fast growing strain. Although many factors could have been responsible for the leg weakness which occurred, it is tempting to hypothesize that the feeding of fish silage products to broiler chickens may enhance their requirements for vitamins and trace minerals.

Organoleptic analyses on the dressed chicken carcasses from the third trial showed that there were no significant differences in the eating quality of the birds fed fish meal, very fresh fish silage, average fish silage or spoiled fish silage.

A three-month feeding trial was carried out with layer birds in which dried silver belly/rice bran silage replaced (on a weight for weight basis) part or all of the silver belly fish meal (61% crude protein) incorporated in the control diet at the 20% level. There was no difference in the results of the silage treatments but the egg production in the fish meal treatment was marginally superior. Organoleptic analyses of the eggs revealed no adverse effects of fish silage feeding, possible effects on egg hatchability were, however, not studied.

The results obtained so far, particularly those of trials comparing fish silage with locally available feed materials are encouraging. However, there are many questions still unanswered and it is perhaps rather premature to make any firm recommendations with regard to the use of dried fish silage production as a feedstuff in poultry rations. Nonetheless, further work is in progress and in view of the economic benefits associated with the use of fish silage it is felt that silage could have a significant role in poultry feeding programmes in Sri Lanka.

5. COMMERCIAL FISH SILAGE PRODUCTION, MARKETING AND UTILIZATION

The economics of producing 300 tons of silage per year using a British Petroleum (BP) silage unit situated in Mannar have been calculated. Using a selling price of Cey.Rs. 5 000/ton (U.S.$ 333) a discounted cash flow analysis has shown that the internal rate of return for 5-year project would be 73%. This indicates that the project would be extremely profitable.

To verify the economic feasibility in practice a 3-month joint venture project has been arranged between IFT in Colombo and a fish meal processor in Mannar, using a BP silage unit made available by the Tropical Products Institute in London. Preliminary trials have shown that the BP unit works very efficiently. Typical by-catch, i.e., silver belly, crabs, puffer fish, sea weed, etc., is comminuted easily and the product is completely liquid within 24 h. This enables the fish which has been ensiled on one day to be mixed with rice bran and spread out for drying on the following day. In the climatic conditions prevailing in the northwest of Sri Lanka the moisture content of the silage/rice bran mixture can be reduced to around 10% in 2-3 days. Currently 1 ton per day of by-catch is being processed into silage using 3.5% formic acid. Rice bran is being added to the liquid silage (1:3) and the mixture sun-dried.

To assess the use of fish silage as a poultry feed under commercial conditions, trials are underway at 5 chicken farms. In Sri Lanka most commercial chicken farms purchase a ready mixed (low quality) poultry mash which they fortify with locally produced fish meal (protein content 40 to 45%). Using this as the reference diet, three other dietary treatments, i.e., a formulated diet containing silage, a formulated diet containing fish meal and a ready mixed mash fortified with silage, are being tested with both layer and broiler birds. Each farm has been consigned sufficient silage to feed up to 1 000 birds per treatment for 3 months. It has been agreed that the farm owners will pay for the silage at the end of the 3-month trial only if they are satisfied with the egg lay/growth rate obtained. If this is the case and the economic feasibility has been proven, the BP unit will be sold to the Mannar firm and the poultry farms will be supplied with fish silage on a regular basis.

6. CONCLUSIONS

Although the trash fish resources available in Sri Lanka are not enormous, they would, if converted to fish silage, have a significant impact on the livestock industry. Technically it is possible to produce stable liquid and/or dried fish silages using prawn trawler by-catch or prawn/fish processing waste. Pilot-scale commercial trials in which 1 ton of fish per day is being processed, are underway. Experimental feeding trials have shown that, compared to alternative locally available protein sources (i.e., fish meal and meat meal), fish silage gives satisfactory results when used in broiler or layer rations at levels up to 20%. However, particularly with fast growing birds the performance of chicken fed silage is not as good as for chicken fed a very good quality fish meal. Although there have been no deaths in any of the trials in which silage has been fed to chicken a leg weakness is sometimes apparent. Further experimental feeding trials are underway to study these problems. A major advantage of fish silage over fish meal is that it appears that silage made from spoiled silver belly has the same nutritional value as silage made from very fresh fish. Although puffer fish are usually found in prawn trawler by-catch, it appears that provided the concentration is fairly low (i.e., less than 1%) they will have little effect on the feeding value of silages made from by-catch. To date, no feeding trials have been carried out in Sri Lanka with animals other than chicken.

The future for fish silage in Sri Lanka looks bright. If all goes well, by the beginning of 1980 silage from trash fish will be produced commercially by one firm in Sri Lanka and it is to be hoped that this will be an incentive for other firms to take up the process. Indeed, a number of requests for information and assistance have already been received. Of particular interest is the fact that one prawn processing firm in Colombo has already produced experimental batches of prawn waste silage. The IFT/VRI team will concentrate their efforts on resolving the remaining problems associated with feeding fish silage to poultry and on the commercial production and utilization of prawn waste silage.

7. REFERENCE

Lantz, A. and S. Gunasekera, Fishery by-products: liquid fish meal. Prog.Rep.Fish.Res.Stn.
1955 Colombo, (1):7-10

DRIED FISH SILAGE AS A PROTEIN SOURCE FOR POULTRY FEEDS FEEDING TRIALS WITH BROILER CHICKENS

K.N.P. Ranaweera, J.A. de S. Siriwardene and H. Manamperi
Veterinary Research Institute
Peradeniya, Sir Lanka

and

K.M. Jayawardena, Q. Gunaratne and A. Villadsen
Institute of Fish Technology
Colombo, Sri Lanka

and

R.G. Poulter
Tropical Products Institute
London, England

Abstract

Dried fish silage made from oil sardine/maize meal (crude protein 26.0 percent) and of silver belly/rice bran (crude protein 32.5 percent) were evaluated as protein supplements for broiler chickens.

In the first trial, oil sardine/maize meal silage was substituted for fish meal (crude protein 51.7 percent) at levels of 50 and 100 percent and the diets made iso-nitrogenous with soybean meal. The substitution did not affect the body weights but significantly improved the feed conversion efficiency.

In the second trial, silver belly/rice bran silage was substituted for meat and bone meal (crude protein 50.0 percent) at levels of 33, 66 and 100 percent and the diets made iso-nitrogenous with soybean meal. The results showed a similar trend to those of the first trial. The substitution did not affect the carcass yield, cooking loss or taste acceptability.

The metabolizable energy content of the oil sardine/maize meal silage biologically assayed was 2 878 Kcal/kg.

The improved feed utilization with the test materials is discussed in relation to the dietary energy content and amino acid profile.

FISH SILAGE FOR PIG PRODUCTION

M. Rangkuti, L. Batubara and M. Agus
Animal Husbandry Research Institute
Bogar, Indonesia

1. INTRODUCTION

In pig and poultry husbandry, feed is the biggest component of production cost and it undoubtedly affects the production system. Supply and price variations in Indonesia are common and often affect the production system. One possible solution is to look for alternative sources of raw material. This can be carried out by utilizing unconventional sources of feedstuffs as well as by increasing the use of locally available feedstuffs.

One feedstuff ingredient, which is used as a feedstuff to provide the essential amino acids for pig and poultry, is fish meal. Fish meal processing requires large amounts of raw material on a continous basis so that it can be produced economically. Therefore, fish meal is often an expensive feedstuff ingredient.

The use of fish silage as a substitute for fish meal has been investigated in many European countries. The method of fish silage production is simple and economic as well as being suitable for small-scale production.

Fish silage production could provide the means to utilize the fisheries waste products or the surplus production that cannot be marketed. During handling, transporting and processing, waste products can amount to as much as 10-25 percent of total production. These unutilized products are high in protein and could form an additional source of income for the fishermen. One means of using this waste fish would be to convert it into animal feed or fish culture feed.

2. FEEDING TRIALS

Research on the use of fish silage as a component of animal feed has yielded conflicting results. Some workers have reported that fish silage gives good results without complications and others report some difficulties in utilizing fish silage (Disney *et al.*, 1977). Accordingly, with a view to introducing fish silage in Indonesia, research on nutritive and biological value should be carried out. The Research Institute for Animal Husbandry, in cooperation with the Research Institute for Fishery Technology, has been carrying out trials on feeding fish silage to pigs. The results of these trials are reported below.

2.1 Trial 1 - To test the tolerance of pigs to fish silage

Liquid silage was made using 3 percent (w/w) of a mixture (1:1) of formic acid (90 percent) and propionic acid (95 percent). The silage was stored for 14 days before being included in the pig diets at levels of 10, 20, 30 and 40 percent. The control diet was made up of rice bran, corn, onggok, soybean, oil cake, mineral and Premix D. The treatments used were as follows:

1. Control diet: rice bran (4 percent); corn (80 percent); onggok (22 percent); soybean oil cake (22 percent); mineral (1.8 percent) and premix D (0.5 percent).
2. Control diet (90 percent) + fish silage (10 percent)
3. Control diet (80 percent) + fish silage (20 percent)
4. Control diet (70 percent) + fish silage (30 percent)
5. Control diet (60 percent) + fish silage (40 percent)

The diets were formulated on an iso-nitrogenous basis, the protein content being approximately 16 percent in each case. The composition of the diets is given below.

Diets	DM (%)	CP (%)	CF (%)	C.Fat (%)	ME[a/] (kCal)
1. Control diet (CD)	86.84	16.08	6.12	2.45	3000
2. CD (90 percent) + FS (10 percent)	84.74	16.83	5.78	2.59	2780
3. CD (80 percent) + FS (20 percent)	77.41	16.76	4.73	2.67	2560
4. CD (70 percent) + FS (30 percent)	70.77	16.56	4.44	2.69	2340
5. CD (60 percent) + FS (40 percent)	67.24	16.37	4.16	2.69	2120

a/ By computation

These rations were fed to five groups of four pigs with an initial body weight of 15-35 kg. The trial lasted for 35 days. The results were analysed by applying a Completely Randomised Block Design (5 x 4) and to compare the average of each treatment the Highly Significant Difference (HSD) technique was used (Steel and Torrie, 1960).

The results of the feeding trial are shown below. They indicate that the inclusion of up to 40 percent fish silage did not result in different dry matter intakes when compared to the control ration. However, a better body weight gain was observed with a 30 percent inclusion of fish silage than with the control. The feed conversion was also better with 20 percent and 30 percent fish silage than with the control ration (2.16:2.24:2.85).

Diets	Feed intake (DM/kg)	Body weight gain (kg/week/head)	Food conversion efficiency
1. Control diet (CD)	7.61	2.66	2.85
2. CD (90 percent) + FS (10 percent)	8.35	3.17	2.70
3. CD (80 percent) + FS (20 percent)	9.63	4.39	2.16
4. CD (70 percent) + FS (30 percent)	10.00	4.49	2.24
5. CD (60 percent) + FS (40 percent)	9.74	4.17	2.36

These results suggest that the application of fish silage up to 30 percent of the animal protein in the ration will produce a good result. The highest conversion observed was obtained with 20 percent fish silage in the ration.

2.2 Trial 2 - Nutritive value compared with fish meal

The fish silage used was the same as in the first trial. The basal diet was again made up of rice bran, corn, soybean oil cake, onggok, mineral and Premix D. The composition of the diets was as follows:

(a) Basal diet: rice bran (25 percent), corn (32 percent), soybean oil cake (15.5 percent) onggok (23 percent), mineral (2 percent), Premix D (0.5 percent)

(b) Treatments: 1. Basal diet (94 percent) + fish meal (6 percent)
2. Basal diet (79.5 percent) + raw fish silage (20.5 percent)
3. Basal diet (80 percent) + boiled fish silage (20 percent), 6 percent FM - 20.5 percent RFS - 20 percent BFS (base on protein)

Raw fish silage and boiled fish silage were given an equi-protein basis with that of the fish meal ration (6 percent), i.e., 20.5 percent and 20.0 percent of the ration, respectively. The rations were given to 24 pigs of 2 breeds (12 heads each) with an initial body weight ranging from 17-28 kg.

A split-plot design was applied (2 x 3 x 4) based on completely randomized block design where breed was the main plot and the ration given was the subplot.

The chemical analysis of the three test diets are shown below.

Diets	DM (%)	CP (%)	CF (%)	C Fat (%)	Ash (%)	NFE (%)
1. Basal diet + FM (6 percent)	69.15	15.32	5.21	5.52	6.67	56.43
2. Basal diet + RFS (20.5 percent)	77.57	15.88	3.79	4.54	5.24	48.12
3. Basal diet + BFS (20 percent)	78.08	15.36	4.00	4.69	5.50	48.53

Note: FM - Fish Meal; RFS - Raw Fish Silage; BFS - Boiled Fish Silage

Date analysis showed that the dry matter intake between the two breeds and among ration treatments, as well as the interaction of breed and ration, did not differ significantly. The same finding was also observed on the 2-weekly body weight gain, that was between breeds (20.84:21.68) and among control ration, raw fish silage and boiled fish silage (7.36; 7.59 and 6.77 kg) and among feed conversion (2.89; 2.89 and 3.31) did not differ significantly. The results of the feeding trials are shown below.

Diets	Feed intake (DM/kg/14 days)	Body weight gain (kg/head/14 days)	Feed conversion efficiency
1. Basal diet + FM (6 percent)	21.15	7.36	2.59
2. Basal diet + RFS (20.5 percent)	21.77	7.59	2.59
3. Basal diet + BFS (20 percent)	20.93	6.77	3.31
1. Breed (A)	20.89	7.18	2.92
2. Breed (B)	21.68	7.30	3.02

Thus, the results of these trials indicate that the nutritive value of fish silage (raw or boiled), when fed to pigs, is equal to fish meal when supplemented at a 6 percent level.

3. CONCLUSIONS

From the results of the two feeding trials, it would appear that liquid fish silage can be used at 20 percent of the ration for growing pigs. Satisfactory weight gain and feed conversion were observed compared with the control diets used.

4. REFERENCES

Disney, J.G., I.N. Tatterson and J. Olley, 1977 Recent developments in fish silage. In Proceedings of the Conference on the handling, processing and marketing of tropical fish. London, Tropical Products Institute, pp. 231-40

Steel, R.G.D. and J.H. Torre, 1960 Principles and procedures of statistics. McGraw-Hill Book Co. New York, Toronto, London

FISH SILAGE AS A FEED FOR FRESH WATER FISH

Hidayat Djajasewaka and Rustami Djajadiredja
Inland Fisheries Research Institute, Bogor, Indonesia

Abstract

The nutritional properties of raw fish silage, boiled fish silage and silage made from boiled fish were determined by assaying growth responses in fresh water fish (common carp). Whole fish meal prepared from the same batch of fresh fish was used as the control. All fish diets contained similar levels of dry fish product, and were made iso-nitrogenous.

Each diet was fed to 64 fish (average weight 100 g) divided into 4 concrete tanks (3 x 1.5 x 0.45 m, with water volume of 2 m^3 and a water flow rate of 0.5 l/sec). The experiments lasted for 77 days and the body weight was measured every 11 days.

Feed was given 3 times a day, the amount being 2% of the total body weight. Body weight gain (g) and feed effeciency were 164, 1.67; 125, 1.73; 131, 1.79 and 187, 1.42 for fish meal, raw fish silage, boiled fish silage and silage made from boiled fish, respectively.

The nutritional value of raw silage and boiled silage were similar to the fish meal control. The nutritional value of the silage made from boiled fish was significantly better than the other silages, including the fish meal control.

1. INTRODUCTION

The nutritional value of fish silage prepared by the addition of 3% of a mixture of formic acid and propionic acid (1:1) as fed to common carp was investigated.

2. MATERIAL AND METHOD

Common carp (hybrid of Taiwan and local breed) with a body weight of 100 g was obtained from a local farmer.

The nutritional value of fish meal (prepared from the same batch of fish as used for the fish silage), silage of boiled fish, boiled fish silage and raw fish silage was determined. Composition and chemical analysis of the diets is shown in Table 1.

Each diet was fed to 64 fish divided into 4 concrete tanks (3 x 1.5 x 0.45 m, with water volume of 2 m^3 and a water flow rate of 0.5 l/sec). Feed was given 3 times per day (at 8.00, 12.00 and 16.00 h), the amount being 2% of the total body weight. The experiment continued for 77 days.

Body weight was recorded every 11 days. Temperature, oxygen content, NH3 and pH of the water were also determined every 11 days.

3. RESULTS AND DISCUSSION

The results of the experiment are summarized in Table 2. There was a difference in body weight gain between the different diets but due to variations between the fish within the treatments, statistically there was no difference between fish meal, boiled fish silage and raw fish silage. However, there was an indication that the body weight gain of fish fed raw or boiled silage was lower than those fed fish meal. This might be partly due to the lack of vitamin B_1 judging by the appearance and behaviour of the fish. This may in part be due to thiaminase which is possibly present in the silage and will cause vitamin B_1 deficiency.

The body weight gain of fish fed silage made from boiled fish was significantly better than that with fish meal or the other silages. This might be partly due to the denaturation of thiaminase during boiling. However, it should also be noted that the texture of the pellet for this diet was better than the other pellets, its water stability being better.

The water pH (6.8–7.0), oxygen content (5.9–8.8 ppm), CO_2 content (5.9–11.3 ppm), DMA (0.8–1.07), NH3 content (0.1–0.3 ppm) and the water temperature (24°C–28°C) were similar in all tanks.

From these experiments it was concluded that fish silage could be used in diets for fresh water fish as a replacement for fish meal.

Table 1

Composition and chemical analysis of the pellet feeds

Ingredient (%)	Pellet A	Pellet B	Pellet C	Pellet D
Fish meal	46.75	-	-	-
Fish silage	-	46.75	-	-
Fish silage (boiled)	-	-	46.75	-
Fish silage (made from boiled fish)	-	-	-	46.75
Banawa (young crab)	5	5	5	5
Wheat flour	28	28	28	28
Rice bran	7	7	7	7
Soybean meal	9.25	9.25	9.25	9.25
Corn meal	2	2	2	2
Vitamin	1	1	1	1
Mineral	1	1	1	1
Chemical analysis (%)				
Moisture	9.16	12.29	15.18	15.69
Crude protein	40.35	40.57	40.60	40.10
Fat	6.26	8.83	9.55	8.85
Ash	13.84	13.99	13.7	13.65
Crude fibre	1.96	1.21	0.69	0.77
Ca	2.33	2.39	2.58	2.34
P	2.17	2.12	2.26	2.11

Table 2

Results of feeding trials

Treatment (for composition see Table 1)	Average weight at stocking (g)	Average weight at harvest (g)	Increase in weight (g)	Feed conversion efficiency
Pellet A	106.41	270.98	164.57	1.67
Pellet B	90.08	215.31	125.23	1.73
Pellet C	98.44	230.31	131.87	1.79
Pellet D	101.33	288.75	187.42	1.42

ECONOMIC ASPECTS

As indicated in the Introduction it was difficult to separate the papers into discrete topics. However, only two papers dealt with economic aspects, as follows:

STUDIES ON THE PREPARATION OF FISH SILAGE ECONOMICS OF PRODUCTION

J. Aagaard, J.G. Disney[1], K.M. Jayawardena and R.G. Poulter[1]
Institute of Fish Technology
Colombo, Sri Lanka

Abstract

The cost of producing a dried silage product in Sri Lanka for use in compound chicken feeds has been evaluated. It was assumed that 2.5 percent formic acid would be used to produce liquid silage from silver belly (*Leiognathus splendens*). Rice bran, in a ratio of 1:3, would be added before sun drying. The yield (% w/w of raw fish) has been taken as 65.3 percent and the final protein content as 32.5 percent.

The capital and oparating costs for a plant producing up to 450 tons of liquid fish silage per year were calculated and a discounted cash flow analysis carried out for 3 levels of revenue, i.e., Cey.Rs. 5 000/ton (U.S.$ 333), Cey.Rs. 4 000/ton and Cey.Rs. 3 700/ton. The internal rates of return were calculated to be 77, 38 and 26 percent respectively for a 5 year project. This indicates that the project would be extremely profitable.

1/ Tropical Products Institute, London, U.K.

FISH SILAGE - ECONOMIC ASPECTS OF PRODUCTION AND UTILIZATION

D. Edwards and J.G. Disney
Tropical Products Institute
London, U.K.

Abstract

In recent years there has been considerable interest in fish silage as a means of utilizing waste fish for animal feed. In the IPFC region particularly, considerable progress has been made on the production of both wet and dried silage products and a number of feeding trials have been conducted. More nutritional trials are needed before commercial production can be recommended and the economic aspects also require further study. Few economic studies have been conducted and the paper presents a cost model for the production of silage from shrimp by-catch in Mexico. The economic viability of the process is considered in relation to the overall policy and strategy of livestock production. It is argued that fish silage production should not be considered in isolation, or by comparison with fish meal, but it should be considered as part of the strategy for overall livestock production. As such, workers in the fisheries field must collaborate closely with their colleagues in the livestock sector. Case studies on the economics of silage production under such circumstances are urgently needed and should be initiated as soon as possible.

1. INTRODUCTION

In small-scale fisheries in the tropics daily and/or seasonal gluts of fish occur and, because of transport difficulties and inadequate processing facilities, these surplus fish are often underutilized. The primary aim should be to utilize such fish for human food but this is often not possible and the production of animal feed represents a possible alternative use. The recent R and D activity in the IPFC region and elsewhere on fish silage is related to the desire to make the maximum use of waste fish and fish offal in situations where the quantity involved, or the transport costs, prohibit conversion into fish meal.

A number of papers have been published on the production and feeding of fish silage. However, only a few have considered the economic aspects (Disney, Tatterson and Olley, 1977; Nicholson, 1976; Gaiger, 1978, and Disney, 1979). The purpose of this paper is to explore the economic aspects associated with the production and utilization of silage by the livestock industry. The Tropical Products Institute (TPI) have programmes on fish silage in the U.K., Sri Lanka and Mexico which are largely designed to assess the technical and economic viability of the British Petroleum Liquid Fish Protein plant (other programmes are in hand on the utilization of poultry and ruminant offal). The results of work in Sri Lanka, including the economic aspects, will be reported upon separately whereas this paper will present cost data relevant to the Mexican situation.

2. THE BENEFITS OF FISH SILAGE

The main advantage of fish silage is that it provides a means of utilizing fish that would otherwise be wasted. The capital investment required is considerably less than that required for a fish meal plant, it is relatively simple to operate and requires a minimum of labour, skilled or otherwise. The major problem of smell associated with fish meal production is not a problem and no energy is required to heat the product, a major factor with the increasing cost of energy.

The disadvantages associated with fish silage relate to the distribution and marketing of the product particularly for a liquid product. This new product will require promotion for it to be accepted as an animal feedstuff. There are also nutritional problems to be resolved particularly the poor growth and palatability of dried products when fed to poultry. Hopefully, the workshop will shed further light on some of these problems.

Subject to the resolution of reported nutritional difficulties, fish silage has considerable potential in fishing industries of the IPFC region. A number of pilot projects, such as the current one in Sri Lanka, are required to test the feasibility of producing and using a silage product on a pilot scale, commercial basis.

3. COSTS OF PRODUCTION

Nicholson (1976) reported that liquid fish silage can compete with fish meal in terms of production costs, mainly as a result of reduced overhead costs resulting from a smaller capital investment. Disney, Tatterson and Olley (1976) also reported that a dried silage product could form the basis of a commercially viable operation in Ghana and Malawi. Disney (1979) also carried out an economic evaluation of the production of fish silage in Sri Lanka (based upon information provided by the IFT and since refined by Dr. Poulter).

In Mexico, Crean and Young (1979) have estimated the costs of production of wet fish silage from shrimp by-catch. These costs are presented in Table 1 as a cost model. The discounted cash flow analysis of the basic cost model assumes a project life of 5 years and a cost of capital of 15 percent gives a selling price of Mex.$ 2 314 per ton (U.S.$ 103 per ton) for the wet fish silage (i.e., the selling price of wet fish silage which would give an internal rate of return of 15 percent over 5 years).

The selling price of the wet fish silage compared very favourably with the cost of alternative sources of protein available for use by the local pig industry. The comparison is shown in Table 2. The effect of various costs of formic acid and an alternative cost of raw material (shrimp by-catch) is also shown.

In terms of the cost per unit of protein the silage process appears to be very profitable. Even when relatively high costs for acid and fish are used the costs of production for silage are lower than those for fish meal. At lower raw material costs (U.S.$ 18/ton) silage is even a cheaper source of protein than soya pressed cake. However, these calculations, while encouraging, do not consider such aspects as the current state of the fish meal industry, the quality of the meal produced, the availability of other feed ingredients and the overall needs of the livestock industry. These aspects are currently under investigation in Mexico.

Table 1

Cost model for the production of 312 tons a year of wet fish silage from shrimp by-catch, Mexico

Item	Cost (Mex.$)	Notes
Capital costs		
1. Land	10 000	
2. Building (shelter)	100 000	
3. Liquid fish protein plant	350 000	Includes allowance for freight and installation
4. Mixer	14 000	Includes allowance for freight
5. Storage drums	12 500	
6. Pick-up truck	150 000	
7. Total fixed capital	636 500	Sum of rows 1-6
8. Working capital	150 000	
9. Total investment	786 500	Sum of rows 7-8

4. ASSESSMENT OF NUTRITIONAL AND ECONOMIC VALUE

The comparison shown in Table 2 indicates one method of assessing the value of the fish silage. This type of comparison, generally made with fish meal, may be on the basis of the cost per unit of protein or the cost per unit of dry matter. However, in the view of the authors this approach has limitations and the analysis can, and should, be extended to obtain a clearer appraisal of the benefits derived from the application of fish silage as an animal feed ingredient.

In Sri Lanka, Jayawardena et al. (1978) have investigated the cost of production of a dried fish silage/rice bran mixture for use as a poultry feed. Disney (1979) has extended the analysis by assessing its value on the basis of a least cost formulation for a poultry feed using materials available in Sri Lanka at the time. Thus the nutritional value of the complete fish silage/rice bran feed mixture and the cost of alternative feed ingredients formed the basis for appraising the production of the fish silage.

A more comprehensive approach is reported by Edwards, Machin and Hector (1979) on the production and utilization of poultry offal silage in the Bahamas. Their work covered (i) field trials to assess the technical feasibility and costs of production of poultry offal silage; (ii) feeding trials with pigs and poultry to determine the nutritional value of the silage and to assess the suitability, cost and availability of ingredients for a balancer ration meal to be fed with the silage; (iii) a financial and economic appraisal of the production and utilization of a pig feed comprising poultry offal silage and balancer ration using a wet feed system. The appraisal showed that this feed compared favourably in terms of nutritional value and cost with conventional imported pig feeds.

Similar, but less comprehensive, work has been reported by Morgan (1976) on the utilization of silage produced from tuna canning offal as an ingredient in pig feed in the Solomon Islands.

In order to adequately appraise the economics of production of fish silage and to identify the potential benefits from its application the wider approach should be adopted. Thus, apart from establishing the costs of fish silage production, this approach necessitates the identification of livestock feed requirements and feeding systems by type of livestock; the identification of potential complementary feed ingredients; feeding trials to establish the acceptability and nutritional value of the fish silage and to identify suitable complementary ingredients for a balancer ration. The total cost of the ration incorporating the fish silage and the balancer meal should be calculated. The cost of this ration and the animal performance should be compared with those resulting from alternative poultry and/or pig feeds. Thus, the acceptability of the product and its implications to the livestock sector in areas such as feeding systems and location should be established.

The status and needs of the livestock industry vary from country to country and within certain countries. In any situation where the introduction of fish silage production is contemplated a thorough review of the animal feed situation should be conducted.

5. CONCLUSION

The thrust of activity within the IPFC region should be toward integrating the work on fish silage production with the needs of the livestock sector. It should take account of demand/supply balances within the animal feeds sector, the overall availability of particular feed ingredients, and the most evident resource constraints in this sector. Thus, in devising a programme of feeding trials to assess the utilization of fish silage these issues have a particular bearing on the feed composition, particularly the ingredients of the balancer ration. Feeding trials should also assess the implications for feeding systems arising from the utilization of the fish silage, in particular the potential for wet feed systems with pigs.

Table 1 (Cont.)

Item	Cost (Mex.$)	Notes
Operating costs		
10. Raw fish	124 800	At Mex.$ 400/ton
11. Acid	117 000	At Mex.$ 12 500/ton(at 3.0%)
12. Thiamine additive	7 020	At Mex.$ 22.5/ton of wet silage
13. Labour	124 800	2 men at Mex.$ 130/day; 1 man Mex.$ 140/day
14. Fuel	22 230	Mex.$ 17.8/gal
15. Electricity	62 400	
16. Water	15 600	
17. Maintenance	32 000	5 percent of fixed capital costs
18. Total annual operating costs	505 850	Sum of rows 10-17

(August 1979 - Mex.$ 50 = £1; Mex.$ 22.5 = U.S.$ 1)

Table 2

Comparison of the cost of wet fish silage with alternative sources of protein, Mexico

Protein source	% Protein	Cost per ton (Mex.$)	Cost per unit of protein (Mex.$)
Soya (pressed cake)	40	6 015	150.38
Fish meal	55	9 130	166.00
	60	10 000	166.67
	65	10 750	165.38
Fish silage			
A[1] (i)[2]	19	2 212	116.42
(ii)[3]	19	2 314	121.80
(iii)[4]	19	2 376	125.05
B[5] (i)[2]	19	2 896	152.42
(ii)[3]	19	2 913	153.32
(iii)[4]	19	2 976	156.63

Notes: 1 - Raw fish at Mex.$ 400/ton
2 - Acid at Mex.$ 9 500/ton (at 2.88%)
3 - Acid at Mex.$ 12 500/ton (at 3.00%)
4 - Acid at Mex.$ 17 500/ton (at 2.50%)
5 - Raw fish at Mex.$ 1 000/ton

6. REFERENCES

Disney, J.G., 1979 Prospects for fish silage in Malaysia, Sri Lanka, Bangladesh and the Philippines, IPFC Occas.Pap., (1979/1):51 p.

Disney, J.G., I.N. Tatterson and J. Olley, 1977 Recent developments in fish silage. In Proceedings of the Conference on the handling, processing and marketing of tropical fish. London, Tropical Products Institute, pp. 231-40

Gaiger, P.J., 1978 Fish silage trials in Hong Kong. Proc.IPFC, 18(3):527-42

Nicholson, R.J.A., 1976 Economic factors affecting fish silage production in the U.K. In Proceedings of the Torry Research Station Symposium on Fish Silage. Aberdeen, Torry Research Station

The following unpublished material has also been referred to:

Crean, K. and R. Young, 1979 Ensilage of shrimp by-catch: Progress report

Edwards, D., D.H. Machin and D.A. Hector, 1979 The production and utilization of poultry offal hydrolysate in the Bahamas, Tropical Products Institute

Jayawardene, M. et al., 1978 Economics in production of silverbelly silage. Report No. 6 on Fish Silage. Institute of Fish Technology, Colombo, Sri Lanka

Morgan, D.J., 1976 The utilization of tuna canning offal for pig feeding in the Solomon Islands. Tropical Products Institute

DISCUSSION

During the workshop there was considerable discussion reflecting the degree of interest in the subject. As is usual in this type of meeting the discussion tended to range from one subject to another and some topics were considered on several occasions. Obviously, in the proceedings it is not possible to record the discussion as it actually happened nor is it possible to cover all the topics in detail. This report of the discussion will largely use the format as proposed by the Chairman, Professor J. Raa of Norway. Thus important items will be considered separately and an attempt will be made to reflect the overall views of the meeting. This will inevitably result in certain points made by individuals being omitted; this is regretted but a full record would be far too lengthy.

Available resources

The basic assumption of all work on fish silage is that it represents a method of improving the overall utilisation of fish resources. The available raw material is therefore important and this subject was considered in some depth during the discussion and in several of the papers presented at the workshop.

The available raw material was summarized by the Chairman as shrimp by-catch, periodic surpluses and by-products of fish processing. By far the greatest source in the IPFC region is the shrimp by-catch and this topic took up the bulk of the discussion, although a periodic surplus may be important at certain locations such as the Bali Strait sardine fishery. Waste from fish processing is probably of greater importance in developed countries.

The main outcome of the discussion on shrimp by-catch was a realization of how little knowledge was actually available on both its extent and composition. The FAO estimate of 3-4 million tons on a world-wide basis is calculated from the total landings of shrimp and assuming a shrimp-to-fish ratio of 1:5. In the opinion of most delegates this was a large underestimate as in the IPFC region the ratio was often as high as 1:20. The quantity also varies from country to country. In Indonesia, Thailand and Malaysia it was considered that massive quantities of by-catch are either discarded at sea or are poorly utilized when landed. In Sri Lanka, Bangladesh and the Philippines the quantities are relatively small.

There was a strong feeling that more detailed information is required on both a national and global basis and the meeting will recommend that more information is collected on the extent of fish waste in the IPFC region.

From slides and photographs it was realized that several species were predominant in the by-catch although the by-catch consisted of many species. In most situations the by-catch has a high proportion of silver belly (*Leiognathus* sp.) and threadfin bream (*Nemipterus* sp.). Goatfish and saurida were also common. However, the relative proportions were considered to vary by location, country and season. It was realized that far more information was required on the composition of shrimp by-catch.

Several papers included data on the composition of by-catch but there is no comprehensive study comparable to the work of P. Snell in Sabah or R. Young in Mexico. More work is urgently required. This work requires month by month sampling on an annual basis and the identification must be conducted by competent biologists to be of real value. It was suggested that the work by Young in Mexico could be used as a model for this work; copies may be obtained from TPI in London. The meeting will recommend that national institutes conduct such studies.

A further aspect of these studies on by-catch is the question of the percentage of juvenile fish in the by-catch. Such fish could grow further and increase the overall level of fish production. This information would be of considerable interest to fish biologists and it was decided to put a formal request to the Standing Committee on Resources, Research and Development of IPFC to commission studies on the composition of shrimp by-catch.

It is not only the species composition of the by-catch which is relevant to work on fish silage. The chemical composition is also important particularly as the level of lipid appears to have an important involvement in the nutritional value of fish silage. Studies should also be initiated on this aspect.

A further aspect on the composition of by-catch received some attention. This is the level of toxic puffer fish in the by-catch. It was revealed that in some countries, e.g., Sri Lanka, the larger puffer fish are discarded at sea but the smaller ones, which make up 0.8 percent of the by-catch, represent a potential toxin when silage is fed to animals. Few countries had any information on the amount of toxic fish present in the by-catch and nothing is known of the effect on the toxin of the silage process. Information is urgently needed as the presence of puffer fish, etc., could provide a serious obstacle to the introduction of fish silage. Work in Sri Lanka on the potential level of toxin in silage, assuming a 0.8 percent content of toxic fish, indicated that this was not a serious problem. However, this needs to be verified and it was agreed that expert advice would be sought from Japan.

To summarize, the discussion illustrated how little was known on the species and chemical composition of shrimp by-catch. National institutes agreed to initiate studies in conjunction with biologists, to provide information on the species composition, seasonal variation and chemical composition of shrimp by-catch. They also agreed to investigate the extent of waste fish in their respective countries.

Production of fish silage

There was a general agreement among the participants that the production of fish silage, at least on a small scale, presents relatively few technical problems, the main difficulties are related to the nutritional value of silage. However, there was considerable discussion on the relative merits of formic acid and propionic acid and the need to incorporate mineral acids as a means of reducing costs.

If liquid silage is stored for any length of time, mould growth occurs at the surface. This must be avoided because toxic aflatoxin could be produced. In Norway and Indonesia equal mixtures of formic acid and propionic acid are used to produce silage and Professor Raa showed by means of slides the control of mould growth by propionic acid. The pH is lowered by the formic acid and the propionic acid then dissociates to give its fungicidal properties. However, it must be emphasized that propionic acid alone should not be used. Work by TPI and in Sri Lanka has concentrated upon using formic acid only. Providing the silage is stirred regularly mould growth need not be a problem and in any case silage for poultry feed is dried within a few days and mould growth is not a problem. After the discussion it seemed clear that if liquid silage is to be stored for more than a few days then propionic acid should be incorporated. If the material is to be consumed quickly or be sun-dried it is not necessary to include propionic acid although it may be wise to include it as a safety precaution. It should be remembered, however, that propionic acid is generally more expensive than formic acid.

There was also some discussion on the level of acid required under laboratory or commercial conditions. It was generally agreed that a safety factor should be included under commercial conditions, particularly in the tropics. In Sri Lanka, for example, 2.5 percent formic acid is sufficient to preserve silver belly under laboratory conditions but an excess (3.5 percent) is used under commercial conditions to guarantee a pH below 4.0 percent despite the additional cost involved. This is particularly necessary as the raw material for silage production will almost certainly be partially spoiled and experiments in Thailand have suggested that spoiled fish require higher levels of acid. There was also a suggestion from Norway that more acid is needed with tropical fish due to their higher buffering capacity. The use of excess acid is also related to the desire to avoid using a pH meter under commercial conditions.

Although the use of organic acid is probably more cost-effective than the heating required to make fish meal there are problems of availability and cost of formic and propionic acid in certain IPFC countries. As these acids are produced as by-products of petroleum manufacture their price has increased in line with oil prices and if cheaper acids could be used it could have considerable advantages. In Thailand for example, formic acid costs three times as much as sulphuric acid and successful silages have been prepared using a ratio of 1:1:0.5 of formic, sulphuric and propionic acids. It was agreed that more work is needed on this subject but a note of caution was made that the preservative action of the organic acids must be maintained particularly during the storage of silage when the pH tends to rise. Also related to the possibility to reduce production costs there was some discussion on the use of locally available products such as tamarind which is cheap and readily available in some countries. Apparently a 5 percent addition of tamarind will lower the pH to 4. It was agreed that this merits some R and D activity.

Two other topics were discussed in this area. Firstly, it was suggested that information is required on monitoring the quality of fish silage. In Sri Lanka measurement of the pH has been found to be a reasonable indication of quality. Secondly, there was a feeling that more work is needed on the production of silage from shrimp waste. Large quantities of this material are available in the region but relatively little work has been carried out on using it for silage production although it is known that much higher levels of acid are required than needed for fish silage.

Lactic acid fermentation

There was considerable interest in lactic acid fermentation as a means of utilizing waste fish, particularly the work carried out in Indonesia, Malaysia and Norway. In the IPFC region fermented foods are well known. The principle of the ensiling process is to add fermentable carbohydrate to fish and the lactic acid produced preserves the product. The meeting was informed that the preservative action is not due solely to the lowered pH but also due to bactericidal and bacteriostatic compounds formed by the lactic acid bacteria. There was some debate about the need to add starter cultures but apparently in many situations the carbohydrate source, e.g., mollasses, contains a natural flora of lactic acid bacteria.

This technology has considerable potential particularly as it should be applicable at the village level. The work in Indonesia has indicated that on a relatively small scale satisfactory products can be made which, when fed to poultry, appear to have fewer nutritional problems than the more conventional acid silage. There was some discussion as to the practical difficulties of introducing this process on a commercial scale but these will be considered later.

In the longer term this approach also offers the prospect of producing a fermented food for human consumption. More research is needed on this subject. In Indonesia, where organic acids are imported and difficult to obtain, it was argued that microbial rather than acid silage offers the best prospect for utilizing waste fish as animal feed.

Areas requiring further investigation on microbial silage were the effect of the process upon fish lipids and whether taint could be a problem. Also the reason why spoilage does not occur despite the fact that it may take two days for the pH to fall below 5 - experience in Norway suggests this is due to the sugar content and the production of bactericides by the lactic acid bacteria.

It was suggested that microbial fermentation might provide a means of using shrimp waste and thus should be tried. Some work in Indonesia on the giant African snail was also reported. Acid silage of this material is difficult but with lactic acid fermentation the pH falls to 3.5 or 4.0 within 3 days and a satisfactory product could be made.

Leading on from this work the meeting was informed of work in Norway which showed considerable promise for the separation of fish components. By manipulating the pH and temperature of fish it is possible to remove fish skin and separate the muscle from the bone; the lipid also separates. This process works by altering the activity of intrinsic enzymes which break down collagen. This approach could be of benefit in silage production since the lipid is separated without emulsifying and can be removed relatively easily.

Animal feeding trials

This subject took up the major share of the discussion as might be expected in view of unresolved problems relating to the feeding of dried fish silage (with say rice bran or maize meal) to poultry. It will not be possible to record all of the points raised but an attempt will be made to present a fair summary.

The discussion centred mainly on poultry feeding but papers were presented on feeding to pigs and fish. From the pig feeding trials reported in Europe there seem to be few problems associated with feeding silage to pigs. This was borne out by the trials in Indonesia as there appeared to be no significant difference between fish meal and fish silage diets. It was also reported that several feeding trials with pigs are underway or planned by TPI in the U.K. and Mexico.

The paper on feeding silage to fish was of great interest as this outlet has considerable potential in the IPFC region. The nutritional value of silage(and boiled silage) was found to be similar to fish meal but the value of silage made from boiled fish was found to be superior to the fish meal control. More work is needed on this subject and the meeting heard of considerable interest in Norway and the U.K. on making fish diets which include fish silage.

The discussion on poultry feeding trials was lengthy and detailed and revolved largely around the trials carried out in Indonesia, Sri Lanka and Thailand with additional information on experience in the U.K. and reports in the literature. Details are given in the various papers and it will only be possible to summarize the discussion but in Indonesia the experiments indicate that when fed high inclusion levels silage has a poor performance when compared to fish meal and problems of "leg slip", perosis, etc., were evident. At lower inclusion levels the weight gain with silage was similar to that obtained with fish meal. Boiling the silage, extracting the oil and adding extra vitamins and minerals improved the performance of the silage but not to the same level as fish meal. The work in Thailand suggested that there was no palatability problem associated with feeding silage and the growth response was only marginally inferior to that obtained with fish meal. In Sri Lanka the findings of several experiments, in which low levels of inclusion have been used, suggest that dried silage gives a poorer growth response than high quality fish meal but is comparable to the poor quality fish meal produced locally. Recent experiments carried out at TPI suggest that at high inclusion levels performance with silage was poorer than with fish meal but at lower inclusion levels the difference was minimal. However, these results do not answer some of the problems observed in earlier experiments at TPI.

During the discussion it became clear that recent experience reported at the meeting did not solve or explain the problems associated with feeding dried silage products to poultry. It seems clear that silage contains some factor which retards the growth of poultry, particularly when fed at relatively high levels. There was considerable discussion on the nature of this effect but no clear conclusions were reached. Discussion centred largely on the freshness of the raw material, the effect of the lipid content and effect of storing the wet silage prior to making the feed. Evidence on the freshness of the material is conflicting; in Sri Lanka this seemed to have little effect but at TPI spoiled silage gave poorer growth rates. There is certainly evidence that removal of the lipid improves the growth response and the age of the silage (or the condition of the lipid) certainly has an effect. For example in Thailand and Sri Lanka the liquid silage is kept for only 3 days, while in Indonesia it is stored for at least 14 days to simulate manufacture at sea.

The real outcome of the discussion was to recommend that further work be carried out on feeding silage products to poultry. Topics to be investigated included storage of the liquid silage and the dried product, to presence of toxic amines, the effect of lipid oxidation (including the use of antioxidants), thiaminase activity and vitamin supplementation and a comparison of fresh and spoiled fish.

One further aspect related to nutritional studies was discussed; this was the question of "taint" in pig or poultry meat due to the incorporation of fish silage in the diet.

Research to date indicated that if the diet contains less than 1 percent of marine lipid on a dry weight basis then problems of taint should not occur. However, this is a relatively low level and in feeding trials arrangements should be made to taste the meat produced. It was also pointed out that the problem is not just related to fishy "taint" but bland flavours have also been reported when fish meal has been fed to poultry and pigs in South Africa.

Commercial application

The meeting was informed of the commercial plants available for fish silage production. A large-scale plant with a capacity of 2-3 000 tons over 200 days per year is operating in northern Norway. It was reported that although this plant is expensive it has economic and environmental advantages over fish meal plants and could have considerable potential in Europe and elsewhere. The British Petroleum plant was also discussed which has a capacity from 300 to 5 000 tons per year, the smallest plant costing approximately U.S.$ 10 000. Several plants are in operation in the U.K., pilot scale trials are underway in Sri Lanka and further trials are proposed in Mexico.

The operation of the BP plant in Sri Lanka was discussed in some detail. The plant is operating well and some 20 tons of dried product are being prepared for large-scale feeding trials by commercial poultry producers. Despite the nutritional problems referred to above it is felt that fish silage can be recommended for feeding to slow growing poultry as a replacement for the locally produced, poor quality fish meal. Provided the commercial feeding trials are successful the plant will be used to prepare a dried silage product on a commercial basis.

A discussion on the commercial application of fish silage indicated the need for further pilot-scale projects in addition to the one in Sri Lanka. It was felt that priority should be given to Thailand although projects on acid and microbial silage should be initiated in several IPFC countries. The technical problems of producing acid silage are few but it was argued that there may be considerable problems in converting the laboratory-scale production of microbial silage to the commercial scale. In the case of both acid and microbial silage it was agreed that there are formidable problems to be overcome in producing silage at sea. It was further agreed that this area had been neglected to date because of the practical problems involved but it was felt strongly that work on the production of silage at sea was urgently needed. Despite the difficulties of vessel design, length of fishing trip and using acids at sea it was felt that work on this subject could be very profitable and Thailand was proposed as the best site for such a project.

Finally, it was felt that fish silage technology has considerable potential in the IPFC region but there would be problems associated with the introduction of a new product. Further studies were needed on the transportation and marketing of both liquid and dried silage products.

Economic aspects

The benefits of the fish silage process over the more conventional fish meal production were discussed. The simplicity of the silage process, the low capital investment, the energy saving and the environmental benefits all argue in its favour. Economic studies carried out in Sri Lanka, Mexico and the U.K. all concluded that fish silage production was a viable alternative to fish meal. This is particularly important in the IPFC region where many fish meal plants are old and produce a poor quality meal; the large throughput required by fish meal plants is also causing difficulties in several countries. Thus, it was felt that the silage process has considerable potential in the region as and when the nutritional difficulties can be resolved.

It was agreed that the production of silage on both a small and large scale should be further investigated. The production and utilization of fish silage should be considered not just as a fish technology problem but more attention should be paid to the needs of the animal feeds industry. It was agreed that case studies should be conducted in several countries of the IPFC region.

Summary

The workshop made a substantial contribution to the state of knowledge on fish silage and reflects the considerable interest in the subject in the IPFC region. Significant progress has been made on the production and utilization of fish silage but inevitably there are many problems needing solution, particularly on the feeding of dried silage products to poultry. More research effort is needed and the workshop helped to define those areas requiring particular attention; these are outlined in the conclusions and recommendations.

To facilitate the research work and coordinate the efforts of the various institutes it was agreed that specific individuals would coordinate the work on specific topics. It was agreed that Wan Johari bin Wan Daud of Malaysia would coordinate the regional studies on by-catch. June Olley from Australia would coordinate the efforts on toxic amines, lipid oxidation, thiaminase, etc. in relation to the nutritional problems and Putu Kompiang of Indonesia would attempt the difficult task of coordinating the work on feeding trials.

CONCLUSIONS AND RECOMMENDATIONS

The number of papers presented, and the quality of the discussion during the workshop, have illustrated the degree of interest in fish silage in the IPFC region. The use of acid or microbial silage techniques as a means of utilizing waste fish are under investigation in most countries in the region. Although outstanding problems remain the increased effort of recent years has produced a wealth of information and pilot-scale commercial trials are being considered. Commercial trials in Sri Lanka, for instance, could commence shortly, subject to the outcome of large-scale feeding trials.

Inevitably the increased research effort has produced as many questions as answers and much more work is required. The emphasis, however, may shift slightly from the solution of production problems to a consideration of commercial application. Many recommendations for future activities arose during the workshop and these are listed below:

1. The papers presented at the workshop represent a major contribution to the field of fish silage. However, a review of the fundamental aspects of the production and utilization of fish silage should also be carried out.

2. More information is required on the availability of waste fish, particularly shrimp by-catch. This will involve assessments by national institutes as well as more widely by FAO.

    (a) Review the extent of waste fish, particularly shrimp by-catch.

    (b) To conduct, on a national and regional basis, studies on the species composition of trawl by-catch using recognized taxonomic names. Also to determine the proximate composition of the whole by-catch and of the predominant species. To be of value such studies should be conducted month by month for at least one year.

    (c) Special attention should be paid to the level of toxic fish in the by-catch. The need to remove these fish must be investigated together with an examination of the effect of the silage process upon the toxic principle. Advice will be sought from experts in Japan.

3. To determine the effect of catching juvenile fish during shrimp fishing the use of selective fishing gear should be considered.

4. Further work should be conducted to investigate the reported decreased rate of growth of poultry when fed dried silage diets. Factors to be studied include:

    (a) Storage of the liquid silage and the dried silage product.

    (b) The presence of toxic amines,i.e., histamine, putrescene and cadaverine.

    (c) The effect of protein degradation.

    (d) Lipid oxidation and the use of antioxidants.

    (e) Thiaminase activity and vitamin supplementation.

    (f) A comparison of fresh and spoiled fish in relation to the above.

    (g) Means of monitoring the quality of fish silage products should also be studied.

    (h) The stability of bacterial toxins under acid conditions.

5. It is recommended that further feeding trials are carried out with poultry, ducks, pigs and fish. Acceptability trials must be conducted by taste panels to examine aspects such as "taint".

6. Related to the production of acid silage more work is required on the combination of mineral and organic acids to minimize costs. Locally available preservatives should also be investigated. Further work on the utilization of shrimp waste should also be carried out.

7. The preparation of silage by lactic acid fermentation shows considerable promise and this methodology may solve some of the nutritional problems associated with acid silage. This method provides the possibility of utilizing waste fish for direct human consumption. More work is needed, particularly on methods of practical application. The possible beneficial effect of the process upon fish lipids should be given special attention.

8. The commercial application of fish silage, including problems of marketing and transportation, should be pursued. There is a need for further pilot-scale projects in addition to the one in Sri Lanka. Priority should be given to the production of fish silage at sea, particularly in Thailand. Pilot projects on the production of acid and microbial silage should be initiated in several IPFC countries.

9. The production and utilization of fish silage should be integrated with local requirements for feeding fish and animals. Conditions vary from country to country but there is an increasing demand for animal feed.

   (a) On the larger scale the utilization of fish silage should fit into the overall animal feed industry. Case studies should be conducted in several countries.

   (b) The production of fish silage on a small scale at the village level should be further investigated. Where appropriate feasibility studies or extension programmes should be initiated.

10. The technology of fish silage is relatively new in the tropics. Acceptance of a new technology will be difficult but in view of the energy saving involved and the reduced enviromental problems, when compared to fish meal production for instance, silage could represent a major advance in fish technology. The meeting also noted that silage technology can also be applied to waste protein from poultry production and animal slaughter.

APPENDIX A

LIST OF PAPERS PRESENTED

1. Possible problems encountered in the application of fish silage in Indonesia
   by R. Moelyanto

2. The status of research on fish silage in Indonesia
   by S. Ilyas and I.P. Kompiang

3. Prospects for fish silage production in Malaysia
   by Wan Johari bin Wan Daud

4. The status of research on fish silage in Malaysia
   by Yeoh Quee Lan

5. The status of fish silage research in the Philippines
   by Susan Canonizado

6. Prospects for the production and utilization of fish silage in Thailand
   by Bung-orn Saisithi and Pong Pen Rattagool

7. Studies on the preparation of fish silage: I. Effect of quality of raw material and type of acid (Abstract only)
   by K.M. Jayawardena, A. Villadsen, Q. Guneratne and R.G. Poulter

8. Studies on the preparation of fish silage: II. Rate of liquefaction of different tissues of silverbelly (Abstract only)
   by K.M. Jayawardena and R.G. Poulter

9. Studies on the preparation of fish silage: III. Dried silage products (Abstract only)
   by K.M. Jayawardena, Q. Guneratne, A. Villadsen and R.G. Poulter

10. Microbial fish silage: Chemical composition, fermentation characteristics and nutritional value
    by I.P. Kompiang, Yushadi and D.C. Creswell

11. Nutritional value of fish silage
    by I.P. Kompiang, A. Darwanto and R. Arifuddin

12. Studies on the nutritive value of fish silage for broiler chickens
    by Pong Pen Rattagool, Niracha Wongchinda, Attaya Kungsuwan and Saifon Swachathamwongratana

13. Fish silage in Thailand: Feeding trials on broiler chicken
    by Pong Pen Rattagool, Niracha Wongchinda and Saifon Swachathamwongratana

14. Fish silage in Thailand: Further feeding trials on broiler chicken
    by Pong Pen Rattagool, Saifon Swachathamwongratana and Niracha Wongchinda

15. Studies on fish silage in Sri Lanka - A summary
    by R.G. Poulter, K.M. Jayawardena, P. Ganegoda and K.N.P. Ranaweera

16. Dried fish silage as a protein source for poultry feeds: Feeding trials with broiler chicken (Abstract only)
    by K.N.P. Ranaweera, J.A. de S. Siriwardene, H. Mamanperi, K.M. Jayawardena, Q. Gunaratne, A. Villadsen and R.G. Poulter

17. Fish silage for pig production
    by L. Batubara, M. Rangkuti and M. Agus

18. Fish silage as a feed for fresh water fish
    by Hidayat Djajasewaka and Rustami Djajadiredja

19. Studies on the preparation of fish drying: Economics of production (Abstract only)
    by J. Aagaard, J.G. Disney, K.M. Jayawardena and R.G. Poulter

20. Fish silage: Economic aspects of production and utilization
    by D. Edwards and J. Disney

APPENDIX B

LIST OF PARTICIPANTS

1.	Mr. David James	Secretary of IPFC Working Party FAO, Rome
2.	Prof. J. Raa	University of Tromsø, Norway
3.	Mr. John Disney	Tropical Products Institute, London, U.K.
4.	Dr. June Olley	CSIRO, Tasmania, Australia
5.	Dr. S.F. Rubbi	Council of Scientific and Industrial Research, Dacca, Bangladesh
6.	Mr. Wan Johari bin Wan Daud	MARDI, Kuala Trengganu, Malaysia
7.	Ms. Mallika Jayawardene	Institute of Fish Technology Colombo, Sri Lanka
8.	Ms. Pong Pen Rattagool	Institute of Fishery Technology, Bangkok, Thailand
9.	Dr. J.L. Sumner	Royal Melbourne Institute of Technology Melbourne, Australia
10.	Dr. M. Salleh Ismail	Dept. of Food Science and Technology Agricultural University, Malaysia
11.	Mr. A. Gildberg	University of Tromsø, Norway
12.	Mr. Hooi Kok Kuang	SEAFDEC, Singapore
13.	Dr. G. Poulter	Institute of Fish Technology Colombo, Sri Lanka
14.	Ms. Susan Canonizado	Bureau of Fisheries and Aquatic Resources, Manila, Philippines
15.	Mr. Sofyan Ilyas	Research Institute of Fishery Technology Jakarta, Indonesia
16.	Mr. Yunizal	)
17.	Mr. Suparno	)
18.	Mr. R. Moelyanto	)
19.	Dr. I.P. Kompiang	Centre for Animal Research and Development Bogor, Indonesia
20.	Mr. Rustami Djajadiredja	Director, Research Institute for Inland Fisheries, Bogor, Indonesia
21.	Mr. Zulkifli Jangkaru	Research Institute for Inland Fisheries, Bogor, Indonesia

22.	Mr. Hidayat Djajasewaka	Research Institute for Inland Fisheries, Bogor, Indonesia
23.	Mr. Leo Batubars	Research Institute of Animal Husbandry, Bogor, Indonesia
24.	Mr. D. Soekirno	Institute of Fisheries for Quality Control Development, Directorate General of Fisheries, Jakarta, Indonesia
25.	Mr. Burhan Lubis	Sub-Directorate of Quality Control, Directorate General of Fisheries, Jakarta, Indonesia
26.	Dr. L.J. Cook	) Centre for Animal Research and Development) Bogor, Indonesia
27.	Dr. J. Roberts	)
28.	Dr. H.T. Ostrwski-Meissner	)
29.	Mr. Pieter H. Meilissa	Maluku Fisheries Extension Service, Ambon
30.	Mr. Rachmat Arifuddin	Res. Institute of Fisheries Technology, Jakarta, Indonesia

APPENDIX C

BIBLIOGRAPHY ON FISH SILAGE AND RELATED TOPICS

(The more important papers are marked thus *)

1. Aagaard, J. *et al.*, Studies on the preparation of fish silage. 4. Economics of production. Bull.Fish.Res.Stn.Sri Lanka, Technology issue (in press)

2. Allsopp, W.H.L., 1977 The utilization of by-catch in shrimp fisheries. In Proceedings of the Conference on the handling, processing and marketing of tropical fish. London, Tropical Products Institute, pp. 287-90

3. *Alvarez, R.J., 1972 A comparative study of the nutritional value of the protein from ensiled fish. Rev.Cubana.Cienc.Agric., 6:219-22 (in English)

4. Anon., 1973 Fish waste goes into the ration. Farmers Weekly Extra - Pig Prod., 18 July

5. ________, 1977 Liquid fish as animal feed. World Fish., 26(7):25

6. ________, 1977 Pig feed from liquid fish. Fish.News Int., 16(7):53

7. ________, 1978 Fishmeal. Developing countries have role in exploiting unused resources. World Fish., 27(2):31-9

8. *Arafudin, R., P. Kompiang and J. Raa, 1978 Preparation of fish silage, its microbial and chemical stability and nutritional value for chicken - a preliminary summary. Paper presented to the Fifth ASEAN Workshop on Solid Substrate Fermentation, ASEAN Sub-Committee on Protein, Bandung, Indonesia, May 1978

9. Austreng, E., A.E. Andersen and A. Skrede, 1979 Syrekonservering av forfisk. Norsk Fiskeoppdrett, 4(1):4-7 (in Norwegian)

10. Backhoff, H.P., 1976 Some chemical changes in fish silage. J.Food.Technol., 11:353-63

11. Barlow, S.M. and I.H. Pike, 1977 The role of fat in fish meal in pig and poultry nutrition. Tech.Bull.Int.Assoc.Fish Meal Manuf.U.K., (4)

12. Bolton, W. and W.A. Dewar, 1969 The absorption of lactate by the fowl. Proc.World Poultry Congr., 12:117-9

13. Canonizado, S.O., 1978 By-products technology and waste utilization. Proc.IPFC, 18(3): 507-14

14. Central Institute of Fisheries Technology, Ernakulam, 1964 Preparation of fish ensilage. Indian Fish Bull., 11(4):20-1

15. ________, 1964 Fishery products of commerce - fish silage. Indian Fish.Bull., 11(3): 12-3

16. Chakraborty, P.K. and P. Madhaven, 1977 A pilot scale plant for the manufacture of fish hydrolysate. Fish.Technol.Soc.Fish.Technol.Ernakulam, 14(2):159-64

17. *Dagbjartsson, B. *et al.*, 1976 A new method for the production of protein hydrolysates from fish viscera and trash fish. Taekni Tidindi, (77):50 p. (in Icelandic)

18. Dean, P. et al., The effect of feeding diets containing white fish meal on acceptability
1971 and flavour intensity of roasted broiler chickens. Can.J.Amin.Sci., 51:15-20

19. *Disney, J.G., Prospects for fish silage in Malaysia, Sri Lanka, Bangladesh and the
1979 Philippines, IPFC Occas.Pap., (1979/1):51 p.

20. *Disney, J.G. and A. Hoffman, A dried fish silage product. In Proceedings of the Torry
1976 Research Station Symposium on Fish Silage. Aberdeen, Torry Research Station

21. *Disney, J.G., I.N. Tatterson and J. Olley, Recent developments in fish silage. In
1977 Proceedings of the Conference on the handling, processing and marketing of tropical fish. London, Tropical Products Institute, pp. 231-40

22. *Disney, J.G., W.H. Parr and D.J. Morgan, Fish silage: preparation, utilization and
1978 prospects for development. Proc.IPFC, 18(3):543-53

23. *Disney, J.G. et al., Development of a fish silage/carbohydrate animal feed for use in
1978 the tropics. Trop.Sci., 20(2):129-44

24. Dreosti, A. and L. Buchanan, Acid preservation of fish. Annu.Rep.Fish.Ind.Res.Inst.
1968 Univ.Cape Town, (22)

25. Dugan, L., Lipids - chemical properties and reactions. In Principles of food science,
1975 edited by O.R. Fennema. New York, Marcel Dekker Inc., Vol.1:166-87

26. Edin, H., Undersøkningar angaande importavstengingens aggviteproblem. Nord.Jord.
1940 brukfoers., 22:142-58 (in Swedish)

27. FAO/IOP, Report of the Joint Mission to plan development of the Sardinella fisheries
1977 in the Bali Strait. Tech.Rep.IOP, (15):51 p.

28. Flores, F.A., The nutritive value of fish silage compared to fish meal. Thesis sub-
1973 mitted to the Department of Fisheries Technology of the College of Fisheries, University of the Philippines

29. Freeman, H.C. and P.L. Hoogland, Processing of cod and haddock viscera. 1. Laboratory
1956 experiments. J.Fish.Res.Board Can., 13(6):869-77

30. Gaiger, P.J., Fish silage trials in Hong Kong. Proc.IPFC, 18(3):527-42
1978

31. Ganegoda, G.A.P., K.M. Jayawardena and R.G. Poulter, Dried fish silage as a protein source for poultry feeds. 2. Effect of fish quality on broiler performance. Tropical Animal Health and Production. Submitted for publication

32. *Gildberg, A. and J. Raa, Properties of a propionic acid/formic acid preserved silage
1977 of cod viscera. J.Sci.Food Agric., 28:647-53

33. Hanson, S.W.F. and J.A. Lovern, The liquid ensilage of fish for animal feedingstuffs.
1951 Fish.News, 26 May, p. 11

34. Hashimoto, K. and N. Fusetani, Toxins occurring in marine organisms important as food
1978 sources in the Indo-Pacific region. Proc.IPFC, 18(3):416-24

35. *Higashi, H. et al., Studies on liquefied fish protein. 1. Nutritive value. Bull.
1965 Tokai Reg.Fish Res.Lab., (43):77

36. *Hillyer, G.M. et al., Evaluation for on-farm use of de-oiled herring silage as a protein
1976 feed for growing pigs. In Proceedings of the Torry Research Station Symposium on Fish Silage. Aberdeen, Torry Research Station

37. Idrus, A.Z., W. Rahimah and J. Abdullah, Fish meal processing; status and prospects in
1978 Malaysia. Proc.IPFC, 18(3):61-72

38. Ishihara, T., Studies on thiaminase. 3. Vitamin B_1 deficiency disease of chicken caused
1974 by inclusion of anchovy in the diet. Bull.Jap.Soc.Sci.Fish., 40(3):309-16

39. James, M.A., K.M. Iyer and M.R. Nair, Comparative study of fish ensilage prepared by
1977 microbial fermentation and formic acid ensilage. In Proceedings of the Conference on the handling, processing and marketing of tropical fish. London, Tropical Products Institute, pp. 273-6

40. Jayawardena, K.M. and R.G. Poulter, Studies on the preparation of fish silage. 2. Rate
1979 of liquefaction in different parts of silverbelly. Bull.Fish.Res.Stn. Sri Lanka, Technology issue (in press)

41. Jayawardena, K.M. et al., Studies on the preparation of fish silage. 1. Effect of
1979a quality of raw material and type of acid. Bull.Fish.Res.Stn.Sri Lanka, Technology issue (in press)

42. ___________, Studies on the preparation of fish silage. 3. Dried silage products.
1979b Bull.Fish.Res.Stn.Sri Lanka, Technology issue (in press)

43. Jensen, J., Fish hydrolysate and fish silage. Final report to DANIDA on the work
1973 carried out on fish hydrolysate and fish silage in the period of May 1972 to April 1973. Copenhagen, Fiskeriminsteriets Forsøgslaboratorium, 22 p. (mimeo)

44. Jensen, J. and W. Schmidtsdorff, Fish silage, low fat and soluble fish protein products.
1977 In Symposium on the Production and Use of Fish Meal, 14-15 September 1977, Szczecin, Poland

45. Jul, M., The protein situation in developed countries versus developing countries.
1977 Proc.Fed.Biochem.Soc., 44(A3):203-16

46. Keay, J.N. and R. Hardy, Fish as food. Part 1. The fisheries resource and its utili-
1978 zation. Process Biochem., 13:2-3

47. Khrishnaswamy, M.A. et al., Nutritional evaluation of an ensiled product for pigs.
1965 Can.J.Biochem., 43:1879-3

48. Kinumaki, T., Possibility of the production and application of liquified fish protein
1978 in the IPFC region. Proc.IPFC, 18(3):492-506

49. *Kompiang, I.P., R. Arifudin and J. Raa, Nutritional value of ensilaged by-catch fish
1979 from Indonesian shrimp trawlers. In Proceedings of the Torry Jubilee Conference, Aberdeen, Torry Research Station

50. Lantz, A. and S. Gunasekera, Fishery by-products: liquid fish meal. Prog.Rep.Fish.
1955 Res.Stn.Colombo, (1):7-10

51. Lisac, H., The liquid ensilage of Mediterranean sardines for animal feeding. Proc.
1961 GFCM, 6:111-9

52. Luscombe, J., Feeding fish silage. Pig Farming, Suppl.Dec.:61-3
1973

53. March, B.E. and J. Biely, The protein nutritive value of liquid herring preparations.
1961 J.Fish.Res.Board Can., 18(1):113-6

54. March, B.E. and R.O. Walker, The effect of shifts in the dietary amino acid pattern
1970 on feed consumption in the chick. Can.J.Physiol.Pharmacol., 48:265-8

55. McBride, J.R., D.R. Idler and R.A. MacLeod, The liquefaction of British Columbia herring
1961 by ensilage, proteolytic enzymes and acid hydrolysis. J.Fish.Res.Board Can., 18(1):93-112

56. Mjelde, A., Melkesyregjaering for a oke holdbarheten av vinterlodde. Meld.SSF Bergen,
1974 1974(3):4-6

57. *Nicholson, R.J.A., Economic factors affecting fish silage production in the U.K. In
1976 Proceedings of the Torry Research Station Symposium on Fish Silage. Aberdeen, Torry Research Station

58. *Nilsson, R. and C. Rydin, Fermentation as a means of preserving organic materials.
1963 Act.Chem.Scand., 17, Suppl.1:174-9

59. *Olley, J., Unconventional sources of fish protein. CSIRO Food Preserv.Q., 32(2):27-32
1972

60. *__________, Abalone silage: an exercise in the conservation of waste. Occas.Pap.
1976 Tasmanian Reg.Lab.CSIRO, (3)

61. *Olley, J., J.E. Ford and A.P. Williams, Nutritional value of fish visceral meals.
1972 J.Sci.Food Agric., 19(5):282-5

62. *Opstvedt, J., Influence of residual lipids on nutritive value of fish. 5. Digestion
1973 and deposition of marine fatty acids in chickens. Acta Agric.Scand., 23(4):217-24

63. *Petersen, H., Acid preservation of fish and fish offal. FAO Fish.Bull., 6(1):18-26
1953

64. Pieris, T.S.S., An attempt to utilize fish waste and trash fish in Sri Lanka. Proc.
1978 IPFC, 18(3):375-9

65. Pike, I.H., The role of fish meal in diets for poultry. Tech.Bull.Int.Assoc.Fish Meal
1975 Manuf.U.K., (3)

66. Potter, D.P., I.N. Tatterson and J. Wignall, Fish by-products: fish meal and fish
1978 silage. Process Biochem., 13(1)

67. Ranaweera, K.N.P., K.M. Jayawardena and R.G. Poulter, Dried fish silage as a protein source for poultry feeds. 3. Feeding trials with layer chicken. Tropical Animal Health and Production. Submitted for publication

68. Ranaweera, K.N.P. et al., Dried fish silage as a protein source for poultry feeds. 1. Feeding trials with broiler chicken. Tropical Animal Health Production. Submitted for publication

69. Ratagool, P., N. Wongchinda and S. Surachathamrongratana, Fish silage production in Thailand. (in press)

70. *Rao, P.D., Ensilage of fish by microbial fermentation. Fish.News Int., 4(3):283-6
1965

71. Saisithi, P. et al., Microbiology and chemistry of fermented fish. J.Food Sci., 31:
1966 105-10

72. Santos, L.M., F. Magno-Orejana and M.N. Bautista, Utilization of by-product of milkfish
1977 processing. Preparation of fish meal and silage. Fish.Res.J.Philipp., 2(1): 56-69

73. Schroder, K. et al., Psycrotropic *Lactobacillus plantarum* from fish and its ability to
1979 produce antibiotic substances. (MS)

74. Sikorski, Z.E., E. Dunajski and S. Koblynski, The reduction of fish in Poland. FAO
1969 Fish.Tech.Pap., (69):17 p.

75. Sinoda, M., P.Y. Lim and S.M. Tan, Preliminary study of trash fish at Kangkar fish
1978 market in Singapore. Bull.Jap.Soc.Sci.Fish., 44(6):595-600

76. *Smith, P. and A.H. Adamson, Pig feeding trials with white fish and herring liquid
1976 protein (fish silage). In Proceedings of the Torry Research Station Symposium on Fish Silage. Aberdeen, Torry Research Station

77. Soliman, H.S. and E.R. Ørskov, Utilisation of fish protein hydrolysate for artificial
1976 rearing of lambs. Proc.Nutr.Soc., 35:91A

78. Sripathy, N.V., Applications of proteolysis for fish utilisation. Seafood Export J.,
1975 7(4):11-8

79. Sripathy, N.V. et al., Fish hydrolysates. 3. Influence of degree of hydrolysis on
1963 nutritive value of liquified fish protein. J.Food.Sci., 28:365-9

80. Stanton, W.R., Microbiological utilisation of agro-industrial wastes in the tropics.
1978 Process Biochem., 13:6-7

81. *Stanton, W.R. and Yeoh Quee Lan, Low salt fermentation method for conserving trash
1977 fish under S.E. Asian conditions. In Proceedings of the Conference on the handling, processing and marketing of tropical fish. London, Tropical Products Institute, pp. 277

82. Stormo, B. and T. Strøm, Ensilering av fiskeslo. Institute of Fishery Technology
1978 Research, Tromso, Norway (in Norwegian)

83. Strøm, T. and B. Eggum, Nutritional value of fish viscera silage. (MS)
1979

84. *Strøm, T., B. Stormo and J. Raa, Fish silage; why not use propionic and formic acid?
1979 In Proceedings of the Torry Jubilee Conference. Aberdeen, Torry Research Station

85. *Summer, J., A feasibility study for fish silage production in the Indo-Pacific Fisheries
1976 Council region. Paper presented to the Second Meeting. IPFC Working Party on Fish Technology and Marketing, Colombo, Sri Lanka, 25-26 October 1976. Rome, FAO, IPFC:FT/76/4:40 p. (mimeo)

86. *__________, Performance of pigs fed on diets containing fish silage: evaluation in
1978 the commercial situation. Proc.IPFC, 18(3):554-8

87. Summer, J. and D. James, Silage production in tropical and subtropical fisheries. In
1977 Proceedings of the Second Annual Tropical and Subtropical Fisheries Technological Conference of the Americas. 17-20 April 1977, Biloxi, Mississippi. College Station, Texas A & M University, Sea Grant College, pp. 59-65

88. Tarky, W., O.P. Agarwala and G.M. Pigott, Protein hydrolysate from fish waste. J.Food
1973 Sci., 38:917-8

89. Tatterson, I.N., The preparation and storage of fish silage. In Proceedings of the
1976 Torry Research Station Symposium on Fish Silage. Aberdeen, Torry Research Station

90. *Tatterson, I.N. and M.L. Windsor, Fish silage. J.Sci.Food Agric., 25:369-79
1974

91. *Tatterson, I.N. and J. Wignall, Alternatives to fish meal. World Fish., 24(5):42
1976

92. Thongthai, C. and M. Siriwongpairat, Changes in the viable bacterial population, pH
1978 and chloride concentration during the first month of Nam Pla (fish sauce) fermentation. J.Sci.Soc.Thailand, 4:73-8

93. *Van Wyk, G.N. et al., Utilization of fish silage: a study on its intake by porkers.
1977 Agroanimalia, 9:13-6

94. *Whittemore, C.T. and A.G. Taylor, Nutritional properties of liquefied herring offal.
1976 In Proceedings of Torry Research Station Symposium on Fish Silage. Aberdeen, Torry Research Station

95. *___________, Nutritive value to the growing pig of deoiled liquefied herring offal
1976 preserved with formic acid (fish silage). J.Sci.Food Agric., 27:239-43

96. Wignall, J. and I. Tatterson, Fish silage. Process Biochem., 11:17-9
1976

97. Wirahadikusumah, S., The effect of fish silage on the quality of hens' eggs and meat
1969 of broilers. Lantbr.-Hogsk.Annlr.35, pp. 823-35

98. ___________, Antimicrobial properties of lactic acid bacteria isolated from fish
1971 silage. Swed.J.Agric.Res., 1:225-7

99. Yanase, M., Studies on liquefied fish protein. 2. Difference in tryptophan content
1965 in enzymatic hydrolysate of fish by different digestion methods. Bull. Tokai Reg.Fish.Res.Lab., 43:87-90

100. Young, R.H., Utilization of by-catch. Inf.Tec.Inst.Tec.Estud.Super.Monterrey, (1978)
1978

APPENDIX D

ANNOTATED BIBLIOGRAPHY

A. AVAILABLE RESOURCES FOR SILAGE

1. By-catch fish

Total world fish catch and the amount available for feed production (7, 46, 2)

By-catch in South-East Asia (21, 85, 19, 87, 75, 22, 64)

By-catch in other parts of the world (100, 96, 44)

What kind of species (21, 75, 100, 30, 49, 13)

Toxic species (100, 34)

Chemical composition of by-catch fish (21, 96, 30, 36, 76)

2. Periodic surpluses of fish

Fish gluts in the Indo-Pacific region (85, 19, 30, 27, 48, 64)

3. By-products of fish processing

Amount of by-products from fish processing in some industrialized countries (96, 94, 20, 29, 84, 74, 48, 61)

Chemical composition (21, 96, 76, 29, 89, 83, 32, 90, 88, 91)

B. PRODUCTION OF FISH SILAGE

1. By acid addition

Commercial production of fish silage today (44, 89)

Preservation effect by different acids (44, 32, 9, 51, 20, 23, 63, 14, 5, 91, 96, 24, 22, 33, 15, 39, 48, 61, 59, 63, 64)

Necessary concentration (21, 49, 89, 9, 51, 20, 23, 26, 63)

Price and availability (85, 87, 82, 22, 63)

Autolysis of fish silage (96, 44, 89, 32, 90, 10, 32, 43, 24, 63)

Chemical degradation of essential amino acids in acid silage (44, 23, 10, 43, 99, 88)

Oxidation of lipids, effect of antioxidants (89, 90, 20, 23, 60, 25, 70, 91, 22)

2. By ammonia

Preservation and solubilization at alkaline conditions (17)

3. By fermentation

Fish sauce production (13, 92, 71, 80)

Preservation by lactic acid bacteria (80, 56, 73, 70, 98, 8, 39, 58, 81)

C. REMOVAL OF OIL

1. Is it necessary (96, 49, 76, 23, 86)
2. Conditions of efficient separation of oil and protein (44, 32, 43, 55)
3. Technology of oil removal from silage (96, 20, 55, 95, 84)

D. ANIMAL FEEDING TRIALS WITH SILAGE DIETS

1. Nutritional value of fish silage compared with other protein feed (19, 96, 44, 30, 49, 36, 76, 83, 51, 20, 26, 43, 86, 95, 70, 52, 4, 88, 77, 65, 54, 6, 38, 97, 3, 22, 47, 35, 93, 28, 8, 53, 61, 79)

2. Can silage be the sole source of animal protein in the feed (21, 19, 44, 30, 43, 86, 70, 77)

3. Any long time adverse effects of feeding silage diets (36, 26, 12, 97, 62)

4. Problems with fishy taint on the animal product (96, 26, 60, 86, 52, 4, 11, 54, 38, 97, 47, 93, 18, 62)

5. Feed value of silage produced from spoiled fish (21, 49, 20, 23, 22)

E. FISH SILAGE OR FISH MEAL?

1. Economic considerations

Production costs (21, 85, 19, 87, 30, 13, 51, 20, 82, 37, 57, 45, 84, 22, 74, 66)
Distribution costs (21, 85, 19, 13, 57)

2. Sociological aspects (21, 87, 13, 45, 84)

F. COMMERCIAL UNITS FOR SILAGE PRODUCTION

Description of equipment for silage production (82, 84)

* * * *

Footnote: The numbers following each subject refer to the abstract numbers in the Bibliography (Appendix C).

APPENDIX E

AUTHOR INDEX

(Alphabetical listing of authors of the papers in these Proceedings giving page numbers).